D0980438

Other titles by Paul A. Johnsgard
available from the University of Nebraska Press

BIRDS OF THE ROCKY MOUNTAINS

DUCKS, GEESE, AND SWANS OF THE WORLD

THIS FRAGILE LAND
A HISTORY OF THE NEBRASKA SANDHILLS

THOSE OF THE GRAY WIND
THE SANDHILL CRANES

# CRANE MUSIC

A Natural History
of American Cranes

Paul A. Johnsgard

University of Nebraska Press • Lincoln and London

⊗ The paper in this book meets the minimum requirements of
American National Standard for Information Sciences—Permanence
of Paper for Printed Library Materials, ANSI Z39.48-1984.

First Bison Books printing: 1998


Library of Congress Cataloging-in-Publication Data
Johnsgard, Paul A.
Crane music: a natural history of American cranes / Paul A.
Johnsgard.
p.   cm.
Originally published: Washington: Smithsonian Institution Press,
1991.
Includes bibliographical references and index.
ISBN 0-8032-7593-5 (pbk.: alk. paper)
1. Sandhill crane.   2. Whooping crane.   3. Cranes (Birds)
I. Title.
[QL696.G84J59   1998]
598.3'2—dc21                                              97-37953
                                                             CIP

None otherwise than cranes 'mongst
clouds together fly,
Dividing air with wings, and echoing
cry for cry . . .

—ANONYMOUS MEDIEVAL VERSE

To Lawrence H. Walkinshaw, George W.
Archibald, and to the memory of Ronald
T. Sauey (1948–1987), all of whom have
understood what cranes are really worth

# CONTENTS

# PREFACE

It has now been almost a decade since I wrote my general monograph *Cranes of the World*. During that time considerable changes have occurred in the world's crane populations and in our knowledge of crane biology. At the time I wrote my book in the late 1970s and early 1980s, for example, only about 100 whooping cranes existed in the wild, and joint Canadian-American efforts to establish a separate population using foster-reared birds in Idaho were just getting under way. At that same time the first important crane sanctuaries were being established on the Platte River, and a variety of privately and federally supported studies on the ecology of the river and its spring sandhill crane populations were being initiated.

Shortly after finishing that book, I was asked to write a general essay on crane biology and some accompanying short accounts of the cranes of the world, to be published in conjunction with a series of life-sized oil paintings of cranes done by the late English artist Philip Rickman, and owned by my friend and fellow crane enthusiast Christopher Marler. Unfortunately, that project never "fledged," and my manuscript was filed away and forgotten for a few years. Then in 1988 an editor from the Smithsonian Institution Press asked me if I would consider doing a semipopular book for them and I suggested that an account of cranes, especially the North American cranes, might make a worthwhile topic. I was already writing a script for a television documentary on sandhill cranes filmed by Thomas Mangelsen, and it seemed to me that an expanded version of that text, along with a comparable one on the whooping crane, plus my earlier unpublished crane manuscript materials might all easily be put together and converted into a short book.

Although several popular books on the whooping crane have appeared during recent years, the sandhill has been essentially neglected. This situation is surprising and unfortunate, in view of the sandhill's much more widespread occurrence and the greater opportunities for the average person to see it. Additionally, the sandhill crane touches special heartstrings for me, for more than any other bird species, I associate it with Nebraska and the Platte River, both of which are very dear to me. Nebraska is my adopted home, and its most important river, the Platte, is my favorite of all the hundreds of rivers of the world that I have seen, from the Amazon to the Yukon. Just as the whooping crane is an endangered species, the Platte is an endangered river, and a supplemental reason for writing this book is to point out once again the inextricable ecological links connecting the cranes, the Platte (and countless other wetlands), and all humanity.

My passion for sandhill cranes began in the spring of 1962, shortly after I had arrived to take up teaching duties at the University of Nebraska after completing postdoctoral studies in England. On a magical Saturday in March, I drove with a graduate student out to the central Platte Valley west of Grand Island, as much to see the spring waterfowl migration as to see sandhill cranes. At that time the cranes had received very little publicity as a birding spectacle, but I felt that I should investigate them nonetheless. It was perhaps just as well that I wasn't emotionally prepared for the sight of countless cranes punctuating the sky from horizon to horizon, or gracefully wheeling about overhead as if they were caught in some ultra-slow-motion whirlwind, their vibrato calls drifting downward like the music of an angelic avian chorus. Not since the days of my boyhood, when I first saw vast migrating spring flocks of snow geese and Canada geese dropping into eastern North Dakota's prairie marshes, was I so completely enthralled, and it was certainly on that particular day of epiphany that I realized that cranes would become as important to my well-being as my beloved waterfowl.

In 1981 I tried to put some of my feelings about sandhill cranes into a small book, *Those of the Gray Wind: The Sandhill Cranes.* That book is as much a story about human attitudes toward nature in general and birds in particular as it is about sandhill cranes. Now, nearly three decades after first seeing sandhill cranes on the Platte, witnessing their return in spring is as much a part of my annual ritual as are Christmas and Thanksgiving, and perhaps even more rewarding. Like Christmas giving, I savor the sandhill cranes of the Platte Valley most completely when I can present them to others as a special gift, and detect in them the same sense of discovery and enormous pleasure that I know so well and feel so deeply. It is for reasons such as this that the present book was written.

In the course of writing this book various people have helped me. My friend and one-time student Tom Mangelsen and I talked for many years about doing a television documentary film on sandhill cranes, and part of the text for this book derives from the wonderful footage that he obtained in the process of completing this film. Additionally, Tom and I have spent (perhaps more accurately, wasted) more time sitting in crane blinds than I care to admit, discussing cranes, waterfowl, and the world in general, and perhaps coming to understand each other better than either of us could ever hope to understand cranes. I must also certainly thank Dr. George Archibald, cofounder and director of the International Crane Foundation, for his good advice and helpful comments on the manuscript. He and our common friend the late Ron Sauey also treated me to the facilities of the International Crane Foundation at various times, and they and their work have provided a rallying point for all crane biologists and conservationists to gather around.

All of these people, and others whom I might equally well have mentioned, have the kind of passionate love for cranes that I can understand, and which I share. I hope that in the course of reading this book a few additional people might develop those same feelings for cranes and their special habitats, which are becoming ever more threatened every year. Since long before medieval times cranes have been considered messengers of the gods, calling annually from on high to remind humans below of the passing years and of their own mortality. Now it is up to humans to take responsibility for controlling our own fate, and also to cry out to protect not only cranes but all the other wonderful creatures that share our fragile earth with us.

# CRANE
# MUSIC

Cranes are the stuff of magic, whose voices penetrate the atmosphere of the world's wilderness areas, from arctic tundra to the South African veld, and whose footprints have been left on the wetlands of the world for the past 60 million years or more. They have served as models for human tribal dances in places as remote as the Aegean, Australia, and Siberia. Whistles made from their wing bones have given courage to Crow and Cheyenne warriors of the North American Great Plains, who ritually blew on them as they rode into battle. These birds' wariness, gregariousness, and regularity of migratory movements have stirred the hearts of people as far back as medieval times and probably long before, and their sagacity and complex social behavior have provided the basis for folklore and myths on several continents. Their large size and humanlike appearance have perhaps been a major reason why we have so often been in awe of cranes, and why we have tended to bestow so many human attributes upon them.

Cranes have provided the basis for a surprising number of

English words that we no longer associate with them. The Greek word for cranes, *geranos* (or *gereunos*), apparently was based on the myth that cranes constantly wage warfare on a tribe of Pygmies, the ruler of whom was named Gerania and had been transformed into a crane by Juno and Diana for neglecting the gods. (A similar myth in India refers to warfare between dwarfs and the fabulous garuda bird.) The geranium plant is so named because of the similarity of the long and pointed seed capsule to a crane's bill. The Romans referred to the cranes as *grues*, apparently from the sound of their calls. The related Latin word *congruere*, meaning to agree, is the basis for the modern English word "congruence," and both derive from the highly coordinated and cooperative behavior typical of cranes. Likewise, "pedigree" is derived from the French *pied de grue*, meaning "foot of a crane," and is based on the characteristic branching pattern of a genealogy. Finally, "hoodwinking" is derived from the practice of sewing shut the eyes of captured cranes in order that they can be more readily tamed and fattened for the pot.

Cranes have been mythically credited with the derivation for several of the letters of the Greek alphabet. Thus, the hero Palamedes supposedly was able to devise several Greek letters simply by watching the convolutions of crane flocks. A similar myth gives the god Mercury credit for inventing the entire Greek alphabet by watching the flights of cranes.

The migratory flights of cranes have probably been observed with interest by humans for millennia, perhaps because cranes generally migrate by day, and also because they typically are organized into coordinated formations during such flights. Edward Topsell (1572–1625), who collected all of the then-available information on the natural history of birds, mammals, and other animals known to the ancient world, wrote at length about crane formations. He believed that the foremost bird in such a formation acted as captain, and that all the subordinate birds of the group organized themselves in such a way as to avoid obscuring its view. Various older birds would reputedly take turns at being the flock leader. Topsell erroneously believed that, should a flock member become tired, it would be supported in flight on the backs, wings, or outstretched legs of other flock members. It has also been widely believed in many cultures that cranes would help transport smaller birds on their migrations by carrying them on their backs.

Various early writers proposed the idea that cranes probably swallowed heavy stones or sand before they began a long flight, with the view that such stones would serve as ballast and prevent the birds from being tossed about by gusts of wind. It was believed that at the end of these flights the birds would cast up the stones or sand. Other writers believed that the stones were carried by the feet, from which they could be easily dropped when they were no longer needed. An equally widely held view was that a flock of cranes would sleep at night only after posting one or more "watch birds," which would stand on one leg and hold a heavy stone in the claws of the other foot. Should such a bird fall asleep, it would drop the stone, thus helping to awaken both itself and the other birds of the flock. This idea gave rise to a Christian morality tale, to the effect that Christians must imitate cranes in their watchful behavior, and avoid falling into sin as a crane avoids falling asleep by holding fast to a heavy stone. With such an anchor, the faithful could find their way through life safely, and upon arrival in heaven the ballast would be turned to gold. Indeed, in heraldry and in the stone carvings of some medieval cathedrals the images of stone-carrying cranes can often be found.

Even more commonly than in the Christian church, cranes have permeated the religions and mythologies of Oriental cultures. A "dance of the white cranes" was performed in China at least as early as 500 BC, and in that country it was generally believed that cranes and dragons transported to heaven those souls that were destined for immortality. It was also believed that old pine trees sometimes were transformed into cranes, or vice versa, both being extremely long-lived. Indeed, in both Chinese and Japanese art there is a recurrent theme of associated pine

trees and cranes, and these icons have generally come to sym-
bolize long life, happiness, steadfastness, and love. Because of
the belief that cranes help support a soul to paradise, a crane-
shaped hairpin may be placed in the hair of a departed woman,
and a representation of a crane may be hung in the window of a
house where there has been a death.

Because of their venerated status, cranes were rarely if ever
killed and eaten in the Orient, although in India they were sometimes sacrificed. In Egypt the birds were captured for food,
together with other waterfowl. Furthermore, in the Temple of Deir-
el-Barari there is a wall painting of captive cranes walking between slaves, with each crane's bill tied down toward its neck,
thus upsetting its balance and making it unable to fly. Other
illustrations of demoiselle cranes in captivity occur in Egyptian
tombs dating from the 5th to the 18th dynasty. Cranes were also
captured and domesticated in ancient Greece, for on a Grecian
vase in the Hermitage Museum at Leningrad, a scene is depicted
of a woman offering a tidbit to a domesticated or captive crane.
At least as early as the late Ice Age in Great Britain cranes were
killed and eaten by humans; British cave deposits of this era have
yielded crane bones, and the bones of a now-extinct crane the
size of a sarus crane have been found in human-associated deposits of the late Pleistocene in Britain and France, of the Neo-
lithic period in Germany, and of the Bronze and Iron ages in
Britain. Because these bones include crane remains of varied
sizes, it has been suggested that perhaps the inhabitants of these
sites may have raised crane chicks for their consumption. At least
as early as the Chou period, some 2,200 years ago, cranes were
raised in captivity by Chinese royalty.

The tales of ancient Greece include many stories of cranes.
For example, it was noted that in Thessaly cranes and storks
would sometimes feed on snakes and thus help to protect the
people there. As a result, the people of that region were forbidden
to kill these birds, a practice that was referred to as *antipalargia*
(from the Greek *palargos*, or stork). Similarly, a mountain on the

Magaris Peninsula was named Gerania (now Yerania), because the people there followed the calls of cranes to higher ground following a flood. The story of the death of Ibycus is even better known; this poet of Rhegium was attacked by robbers and mortally wounded. As he lay dying he saw a flock of migrating cranes overhead, and with his last breath told the robbers that the cranes had seen his murder and would avenge his death. Later, in the Corinth market one of the robbers happened to see a flock of cranes overhead and called out in fear to his friends, "Behold the cranes of Ibycus!" On being overheard, the men were questioned and arrested, and later confessed to the murder of Ibycus.

In a somewhat similar fashion, the sighting of cranes has been associated with death in various other cultures. For example, slaves of the American South believed that if a crane should circle over a house three times, somebody in that house would soon die. An ancient counterpart of this belief may be Pliny's story that the oldest of a flock of cranes would fly around in a circle three times before the flock was due to leave on migration, and then fall down and die of exhaustion. Perhaps these and similar stories derive from the fact that prior to migration cranes do indeed spend much time circling in thermals on sunny days, and ride thermals to great heights immediately prior to setting out on long migratory journeys.

The actual migratory journeys of cranes are no less interesting than they were imagined to be by the peoples of medieval times. In recent years it has been possible to follow these movements very closely, by using radar or radiotelemetric devices or by following migrating flocks in small airplanes. It is now known, for example, that Eurasian cranes, and probably most other cranes as well, maximally utilize their soaring abilities during migration by exploiting the lifting potential of thermal winds, and then gliding in close formation for great distances while seeking out another thermal. Eurasian cranes may thus soar to heights of more than 6,500 feet while in thermals, and their thermal-assisted climbing abilities are especially valuable between about

1,500 and 5,000 feet. Using radiotelemetry, it has been found that greater sandhill cranes can fly nonstop as far as 360 miles during a 9.5-hour period, averaging some 38 miles per hour. This generally agrees with estimated air speeds for Eurasian cranes of from 37 to 52 miles per hour. Observations on migrating whooping cranes indicate that similar daily migration patterns occur, with single-day trips of up to 510 miles reported, but with most daily movements of less than 200 miles and lasting about six and one-half hours.

In the case of the sandhill crane the birds prefer to fly on clear to only partly cloudy days. They normally land before dark, and usually begin to arrive at roosting sites by about sundown. Nearly all migratory flight in this species occurs at elevations below 6,000 feet, generally between about 1,000 and 3,000 feet. Such altitudes are high enough that landmarks are visible from great distances, and place the birds well above ground turbulence or obstacles. The birds also choose those days for migratory flights when they can exploit following winds rather than face crosswinds or headwinds. On the rare occasions when sandhill cranes have been observed migrating during inclement weather, barometric pressures have been rising in those areas toward which the birds were flying. Equally remarkably, sandhill cranes have been observed to terminate a migration leg early in the day, apparently sensing the approach of bad weather well before it has actually arrived.

Flock sizes of migrating cranes vary greatly, probably influenced by such factors as total population size, levels of social tolerance or gregariousness in the species, degrees of disturbance on roosting and foraging areas, and time of year. In whooping cranes, for example, spring flocks average somewhat larger than fall flocks, but in both seasons the average flock sizes are rather small, 3.1 and 2.6 birds respectively. Flock sizes of field-foraging sandhill cranes during spring migration in Nebraska are also typically small. Over three-fourths of such flocks have no more than 50 birds, with the most common social units of two or three birds probably representing individual pairs or family groups. However, the roosting flocks are much larger. At times these huge flocks of birds standing in the safety of the shallow river water number 15,000 or more. These large numbers reflect the relatively few areas of the Platte River that still comprise ideal roosting habitats, and the resultant crowding of birds into these confined stretches of river.

The main factor affecting the daily timing of roosting flights is light level, with the majority of the birds arriving at the roost by sunset, nearly all of them within 15 minutes after sunset. Delayed returns to the roost most often occur under conditions of a clear sky, moderate to high temperatures, and no wind. Similarly, morning departures from roosts are associated with sunrise; more than half of the birds usually leave the roost by a half-hour after sunrise, and nearly all will have left during the first hour. However, heavy clouds, fog, rain, and strong winds all tend to delay morning departures. Judging from limited observations on other crane species, much the same pattern of diurnal activity seems to be typical of cranes in general.

Cranes take flight from a running start into the wind, finally springing into the air and slowly gaining altitude. In flight they present an appearance distinctly different from that of geese, in that the wingbeat is more shallow and the upstroke is noticeably more rapid than the downstroke. This rapid upstroke is especially conspicuous among frightened birds trying to gain altitude quickly. Furthermore, perhaps because of their less labored flight than that of generally heavier birds such as geese or swans, they rarely maintain a fixed formation for any length of time, except when migrating at high altitudes. Instead, the flock pattern is constantly undulating and changing, without any definite lead bird for much of the time. At fairly close range the long and trailing legs of cranes also visually set them apart from geese, although during cold weather it is not uncommon for some of the flock members, especially young birds, to tuck their legs forward into their flank feathers and thus assume a surprisingly gooselike profile. Landing is also done against the wind, with the legs dangled pendulum-like, providing for a lowered center of gravity and increased stability, as the tail is spread and the wings cupped. In this way the birds descend parachute-like almost vertically to their roost, finally breaking their descent during the last few seconds by wing flapping.

While flying, and especially during landings and takeoffs, cranes utter a constant clamoring, enabling pair and family members to maintain vocal contact amid the confusion of flock movements. Although it remains to be proven, there seems little doubt that cranes must be able to recognize their mates or other family members by their vocal traits, for it is common for pairs to

Unison call of the
greater sandhill crane.

maintain "conversational" contact with one another when they are out of each other's sight. When lone birds have somehow been separated from their social groups, it is common to see them flying back and forth over roosting flocks, calling almost constantly.

Indeed, it is the mutual calling of paired cranes that provides a basic key to the understanding of their social bonding, and no observer of cranes can begin to understand their interactions without some appreciation of the importance of this mutual calling behavior. Although there are several other contact calls in cranes, the most important of these is the "unison call." This call, which develops during the second or third year of life, is a complex and extended series of notes uttered by paired birds. It is uttered in a time-coordinated sequence, with the birds typically also standing in a distinctive posture and a specific spatial relationship to one another. This posture is always an erect and alert one, with the wings folded and the primaries often drooped while the inner elongated and ornamental "tertial" feathers are raised. The birds are oriented side by side or facing one another. The "unison" call sequence may last from only a few seconds to a minute or more. The associated vocalizations are typically the loudest and most penetrating of any of the species' calls. In some species, such as the African crowned cranes (genus *Balearica*), the throat is variably expanded to serve as a resonating chamber, while in most of the species of the genus *Grus* the variably elongated trachea evidently serves a similar role as the primary resonating agent.

Although there are considerable differences among crane species, it is usual for a particular sex to initiate the call; very shortly the other sex joins in, often calling in counterpoint to its mate. Most often the female utters shorter calls, and in the genus *Grus* the female usually utters two calls for every call of the male. However, in the more "primitive" genera of cranes, the call notes tend to be fairly short in duration, and both sexes have calls of about the same length. In most of these non-*Grus* cranes, the trachea is relatively short, and the calls typically lack resonance and associated harmonic development. Just as in various swans such as trumpeter and tundra swans, increased tracheal length in cranes is correlated with increased volume and resonance of their primary vocalizations, although the exact means by which this acoustic effect is achieved is disputed. The crane species with the loudest and most penetrating unison calls (whooping crane, Japanese crane, and sarus crane) are those that combine relatively large size, conspicuous plumage patterning, high levels of territoriality, and low social tolerance during nesting, in addition to having greatly elongated tracheal structures. Similar kinds of correlations apply to the swans having variably elongated tracheal structures.

These observations and correlations suggest that the unison call has a variety of social functions. Perhaps most importantly, it seems to be a basic mechanism for individual pair bonding and pair maintenance, and at least in some species it may also serve as an important sex-recognition device. It also serves as a territorial advertisement call and as a general threat call, as it is often stimulated by the intrusion of a potential enemy into the breeding area of an established pair. It might also serve as a synchronizing mechanism for the pair members, helping to bring them into reproductive condition simultaneously. Recent research on sandhill cranes tends to confirm this idea. Males are evidently brought into reproductive condition by increasing day length; ovarian development in females evidently requires further stimulation through unison calling behavior with their mates.

A variety of other call types occur in most cranes; studies by Dr. George Archibald suggest that in the genus *Grus* there are three calls that are uttered only by chicks, as compared with eight that are characteristic of adults. In species that have especially loud and penetrating adult voices, the young birds go through a "voice break" period near the end of their first year,

14

when the high-pitched peeping voices of the young are replaced by much lower and more guttural calls. No apparent anatomical changes in the tracheal length or structure occur at this age, and so the physiological mechanism for this vocal change remains obscure.

Besides their call similarities, cranes share a number of "egocentric" or individualistic behavior patterns that are fairly uniform throughout the entire group. All cranes older than chicks usually sleep while standing, often with one leg raised. However, both young and adult birds may sometimes rest or sleep in a

Lesser sandhill crane.

sitting posture, with the legs folded underneath and the abdomen resting on the substrate. This is also of course the incubating posture, with the head resting comfortably on the breast, or lowered almost to the ground when the incubating bird is trying to avoid detection. While sleeping, the birds often tuck their bills into their scapulars, but during incubation they rarely if ever sleep, and the incubating bird seems constantly alert to danger. Occasionally, a standing crane may rest on its "heels," although this posture is not nearly so common as it is among some heavier long-legged birds such as storks.

Like nearly all birds, cranes drink by quickly dipping their bills, then tilting them upward to swallow. Seeds and insects they obtain from the substrate by pecking, or by digging and probing with the tip of the bill. Prior to taking off, cranes often assume an "intention" posture in which the head and extended neck are gradually lowered to an almost horizontal position. This distinctive posture may help to coordinate flight in a group, or at least warn others nearby that one of their flock members is about to take off. Cranes also perform several stretching behaviors. These include the simultaneous stretching of one wing and the leg on the same side of the body, a double wing stretch about the back (with the head and neck simultaneously lowered and stretched forward), and a similar double wing stretch but without the head lowering.

All cranes preen in a consistent manner. This involves extensive nibbling, drawing of the feathers through the slightly opened bill, and associated oiling behavior. In a few species of cranes (especially sandhill cranes, and to a lesser degree Eurasian cranes) there is an interesting related behavior that is performed by breeding adults immediately prior to nesting. In sandhill cranes the birds "paint" almost their entire body plumage with mud or rotting vegetation that they probe for with their bills, gradually staining their basically grayish plumage into a brownish to reddish brown color. In this way the entire bird eventually assumes an appearance that closely matches the color of their dead-grass nesting substrate. During fall and winter these stained feathers are gradually lost by molting, and so the process must be repeated annually. The Eurasian crane paints itself less extensively than does the sandhill, and probably hooded and black-necked cranes paint to some degree. Only one essentially white-plumaged crane, the Siberian, is known to exhibit similar behavior, and in this species the staining is mostly limited to the lower neck region, often producing a saddle-like pattern.

Preening, feather ruffling, and similar body care activities also occur in a social or "display" context in most and probably all cranes. These movements often scarcely differ from their normal nondisplay counterparts performed in nonsocial situations, and are usually overlooked by the casual observer. However, they are among the most important of the visual signals of cranes, and careful watching will often allow the observer to gain a keen insight into the social interactions of a crane flock, as well as judge crane responses to humans or other animals.

The commonest social displays of cranes in aggressive situations consist of "ritualized" preening movements, which may seemingly be directed toward the back, wings, or elsewhere on the body. These preening movements are performed silently, but the preening bird rarely if ever takes its eyes off the individual toward whom the display is actually directed. Sometimes the preening is interspersed with feather ruffling on the back, inner wing feathers, or the body feathers in general, and commonly the bare head skin is maximally exposed and intensely red during such behavior. A stiff-legged, marchlike approach is often part of the aggressive repertoire, usually with the bill downward so that the bare crown is directed toward the opponent. In some cranes the bird may even sink to the ground in a "crouch-threat," with the wings somewhat spread and the bare crown skin greatly expanded. A very common aggressive display is to spread or variably lower both wings while facing the opponent, especially for those birds that are defending their nest site or young. This display sometimes grades into "broken wing" behavior, in which the displaying bird attempts to decoy the intruder away from the immediate vicinity.

Of all the social displays of cranes, none is more interesting or complex than "dancing." Dancing behavior has been observed in all crane species, but it has been carefully studied in only a few, and its functions are still both controversial and perhaps manifold. Although dancing varies greatly in speed and intensity among cranes (smaller species such as the demoiselle dancing with greater speed and vigor than larger ones), it seems to have a few common components in all. The two major ones consist of

a lowering of the head nearly to the ground while simultaneously lifting and spreading the wings, and a sudden return of the head upward and downstroking of the wings. Actual jumping often accompanies this return phase, in which the bird may also pick up and throw upward a stick or piece of vegetation. Sometimes two birds perform these activities in synchrony or near synchrony

while facing one another or standing side by side. At other times a single bird may dance, or a large group may participate in the activity to varying degrees. The behavior tends to be contagious, and a period of intense dancing may quickly spread through a social group. Dancing sometimes leads to flying, especially when it is stimulated by a threatening outside source, such as the appearance of a potential predator. Dancing also sometimes leads to fighting among the participants. It can occur at any time of year, and among birds of all age groups. It is prevalent among birds that are forming or have recently formed pair bonds, and so its possible role in this general process cannot be dismissed. However, dancing often seems to reflect a general sense of excitement or limited aggression among cranes, and as such it has less circumscribed functions than courtship.

Whatever the possible roles of dancing, it is not a direct prelude to fertilization. Copulation is apparently rather infrequent in cranes and is limited to the period immediately associated with egg laying. Judging from several published descriptions, copulatory behavior can be initiated by either sex, but most often is begun by the male. He walks toward his mate in a distinctive "parade-march" posture with his tertial feathers raised, his crown expanded, and his bill pointed upward, all of which presents an appearance of dominant hostility. The female, if receptive, responds by spreading her wings somewhat and holding her neck and body lowered into a somewhat diagonal posture. This provides a rather flattened surface on which the male is able to stand, after first leaping onto her back. He bends his toes around the leading edge of the female's wings, and flaps his wings to maintain his balance while trying to achieve cloacal contact. Following copulation, both sexes usually participate in a postcopulatory display that may last up to about 20 seconds. This consists of various forms of ritualized threat display, such as feather ruffling, preening, and crown expansion. In the Japanese crane, an elaborate mutual bowing and neck arching are invariable postcopulatory displays.

Cranes are unusually long-lived birds, and tend to place their nests each year in the same or nearly the same location as they did the year before. Based on studies of color-banded whooping cranes, it appears that experienced pairs return to the same "composite nesting area" (the collective territories used in past years) to reestablish their nesting sites, and that other birds are unable to force resident pairs from these areas. Nests are constructed of herbaceous vegetation and are usually fairly close to water or at times even surrounded by shallow water. Nests that are built in water tend to be more bulky than those constructed on land, and both of the species that nest on dry land (blue and demoiselle cranes) construct little or no actual nests, placing their eggs in a shallow depression that may at most contain a few straws or pebbles. In all cranes, incubation begins with the laying of the first egg, and both sexes participate fairly equally in it. The eggs are attended from the beginning of incubation until hatching occurs, which varies from as few as 27 days to as many as 40 days in different species, averaging 31–33 in most.

Since crane eggs are incubated from the time they are laid, and since they are usually laid about two days apart, the young typically hatch on different days. Usually the first-hatched chick will remain on the nest for much of its first day of posthatching life, although it may leave the nest for a short time and be tended by the nonincubating parent (typically the male at this period) while the other continues to incubate. Except in the crowned cranes, two eggs constitute the maximum clutch (wattled cranes frequently have single-egg clutches, and crowned cranes may have as many as four). The nest is typically abandoned soon after the hatching of the second egg, or even after the first egg has hatched, in the case of wattled cranes. Shortly after the last young has hatched, the chicks are led away from the nest site by both parents, often into heavier cover.

During the long fledging (preflying) period of from about 60 to 150 days, the young birds gradually acquire their juvenile plumage, while at the same time their parents typically undergo their annual molt. Adults of most species of cranes lose their major flight feathers rather abruptly during this postbreeding molting period and thus become flightless for a time. In a few species (at least the demoiselle and the crowned cranes), however, there is no actual flightless period, since the period of wing molt is much more prolonged. In yet other cranes there seem to be age-related and possibly also individual variations in the pattern of wing molting and thus the degree of flightlessness. Additionally, it has been suggested that some cranes may molt their flight feathers only on alternate years. In the case of the sandhill crane it is now evident that second-year subadult birds usually do not replace any of their flight feathers except for a few inner secondaries, and thus do not become flightless at all during that year. They also retain some of their juvenal wing coverts, which are buffy-tipped, incidentally providing for in-hand recognition of birds representing this age class. During their third year all of the secondaries and a few inner primaries are replaced, but the outer juvenal primaries are still retained. Buffy-tipped primary coverts are only rarely present in this age class. Finally, after their third year, there is an irregular molting pattern of wing feathers, and these older birds thus exhibit intermixed worn and unworn primaries representing differing molt periods and feather generations.

Because of the very long incubation and fledging periods of cranes, not surprisingly, only one breeding cycle is completed per year. However, there may well be a still unknown incidence of renesting by pairs that fail to hatch their first clutch, at least in those species or races living in fairly temperate climates. For example, the Florida race of the sandhill crane exhibits a seven-month spread of egg records, whereas the Alaskan population exhibits only a four-month spread. The mostly tropical wattled crane of Africa has a twelve-month spread, with a poorly defined peak of egg records between May and August.

Not only do cranes have a single reproductive effort each year, but they also exhibit a pattern of deferred reproductive

Greater sandhill crane chick and egg.

maturity. There is little information on the average age of initial breeding among most wild cranes, but among cranes raised in captivity, it is unusual for breeding to occur less than four years after hatching. The average for 11 males and 14 females of various crane species raised in captivity was slightly over seven years, with males averaging slightly sooner than females. However, the artificial conditions and stresses of captivity are probably not a close reflection of the situation in nature, and thus it is probable that individuals of at least some species regularly breed in their third year of life. Studies on wild whooping cranes indicate that two three-year-old males attempted to breed unsuccessfully, but fathered chicks the following year. One female produced chicks when five years old, and another at five or six years. In another study involving 18 known-age Florida sandhill crane pairs, the youngest two pairs were leading young in the fall at five years old. Among 28 pairs of greater sandhill cranes, on the other hand, the youngest pairs leading young in the fall were three years old; 4 of 17 pairs of this age class were observed with young. Thus there may be racial differences in the attainment of reproductive maturity in this species, but this is still unproven.

Given the small clutch size and the apparently low hatching and fledging success rates of cranes, it is not surprising that they have perhaps the lowest recruitment rate of all birds that are legally considered game birds in North America. This statistic represents the proportion of young birds in the population each fall, and is an important measure of reproductive success. It is unusual for more than about 10 percent of most wild crane populations to be comprised of first-year birds, which means that the adult mortality rate cannot rise above 10 percent per year for prolonged periods without significant adverse effects on populations. Surprisingly little is known of relative age structures in most wild crane populations, in spite of the fact that several of these are rare or endangered, and that it is usually fairly easy to determine visually the incidence of young birds in wild flocks. This is one area where amateur bird watchers could provide extremely valuable information for conservation agencies. By monitoring yearly and long-term changes in age ratios of cranes, an indirect but accurate measurement of the past season's breeding success and the long-term outlook for the species could be easily obtained. In the case of greater sandhill cranes, the incidence of young in populations has recently ranged from about 5 percent (in the declining Central Valley population of California) to about 14 percent (in the expanding Rocky Mountain and eastern populations). Florida sandhills have also shown recruitment rates of 15 percent recently, while lesser sandhills have averaged between 7 and 12 percent and Mississippi sandhills only about 2 percent. Recruitment rates in the Wood Buffalo–Aransas flock of whooping cranes averaged about 11 percent during the two-decade period between 1965 and 1985, and about 15 percent between 1945 and 1965.

So far, too few cranes of most species have been banded to be very certain of their normal mortality rates. Observations on an unhunted and color-banded population of Florida sandhill cranes indicate that among those individuals that were banded as nonjuveniles, about 25 percent survived for at least 7–8 additional years, and about 10 percent of them were still alive 9–10 years after banding. These observations would suggest that an approximate 85 percent survival rate (or a 15 percent annual mortality rate) might occur in protected postjuvenile sandhill crane populations.

Annual survival rates for the two decades between 1965 and 1985 among whooping cranes that have been surveyed annually at Aransas (thus excluding all those juveniles that never lived long enough to reach Texas) have averaged about 92 percent, as compared with about 87 percent during the two decades prior to 1965. Survival rates among juveniles are substantially lower. For example, estimated survival rates of chicks from hatching to fledging at Wood Buffalo Park have ranged from only about 25 percent to as high as 86 percent in various years, and approximately three-fourths of the newly fledged birds survive their first fall migration to Aransas National Wildlife Refuge. Overall survival from the time of banding of unfledged whooper chicks through the following twelve months of life is probably about 60–65 percent. Thus, of all the whooping crane chicks that hatch at Wood Buffalo Park, probably no more than a third to a quarter of them survive long enough to begin breeding at about four or five years of age.

Another important aspect of crane biology about which little is known concerns their effective length of reproductive life. Do cranes remain reproductively active for as long as they live, and, if so, what is the maximum longevity of wild cranes? It is well known that at least in captivity cranes are among the longest-lived of all birds. There are many instances of cranes known to have survived more than 40 years in captivity, and even one case of a male Siberian crane that was believed to be 83 years old when he recently died. Until his late 70s he was still reproductively active, fathering three offspring at an estimated age of 77! Another male white-naped crane at the International Crane Foundation is more than 50 years old and is one of the best semen producers (for artificial insemination) there. What is known of wild crane mortality rates would suggest that few birds are likely to survive beyond about 25 years under natural conditions, and thus it is highly unlikely that any cranes die of old age in the wild, or become infertile because of advancing age.

Of the 14 or 15 species of cranes in the world (the crowned cranes are regarded by some authorities as one species and by others as two), two have been classified as endangered and threatened with possible extinction by international agencies such as the International Council for Bird Preservation (ICBP) and the International Union for Conservation of Nature and Natural Resources (IUCN). These are the whooping crane and the Siberian crane. Three other species (Japanese, hooded, and white-naped) are classified as "vulnerable," and one species (the black-necked) has been classified as of "indeterminate" status. Recent counts of the black-necked crane indicate that its status should be changed to "endangered," while perhaps the Siberian crane can be safely shifted to the "vulnerable" category. Besides the whooping, Siberian, and black-necked cranes, other cranes with probable total world populations of 5,000 or fewer and probably vulnerable to extinction include the white-naped, Japanese, wattled, and perhaps the blue crane. Thus more than a third of the world's crane species can be regarded as vulnerable or endangered, and few if any are totally safe.

Humans have attempted to preserve the rarer species of cranes in various ways, such as establishing sanctuaries in locations critical to their breeding or wintering and trying to protect them from hunting along their migratory routes. However, the problems are compounded by the long and frequently international routes that the birds take, often requiring the cooperation of several different nations for protection of habitats along the route. Some of these nations are more developed than others, and they have differing degrees of conservation orientation. Even in the more highly developed ones, it is ultimately up to the individual to recognize and refrain from shooting or disturbing these sensitive birds. Two of the world's rarest cranes, the whooping crane and Siberian crane, provide examples of the problems of protecting such species from extinction. After nearly a half-century of total protection and intensive conservation activities, the whooping crane is making a slow but determined comeback from what appeared to be certain extinction in the late 1930s. By comparison, the first serious conservation efforts on behalf of the Siberian crane began in the USSR and India only about a decade ago, when fewer than 200 birds were thought to exist. The discovery in 1980 (after two years of intensive searching by Chinese biologists) of an additional wintering flock of about 100 Siberian cranes on the lower Yangtze River in Jiangxi Province of eastern China was thus of great importance. In 1981 about 230 of the birds were present at that locality, along with substantial numbers of white-naped and hooded cranes. Since then the size of that wintering flock has increased to more than 2,000 Siberian cranes in only a relatively few years, bringing hope that this "lily of birds" can soon be officially removed from the list of endangered species.

When we are considering the costs and benefits of saving endangered species, it might be worth remembering that cranes are among the oldest of living bird groups, and the sandhill crane in particular is the oldest known currently existing bird species, based on fossil remains attributed to this species that are about nine million years old. Cranes were already on the scene when the earliest primates were small, hesitant, and shrewlike creatures that might well have cowered in their own shadows in fear of being eaten. Cranes witnessed the genesis of many of the major river systems of the world, and the formation of the tundras, prairies, and savannas that many of them now call home. Indeed,

27

a few million years ago the cranes of eastern Africa were perhaps startled when some hairy-backed creatures from the nearby forests wandered out onto the savannas and awkwardly began to harvest the very seeds and insects that they themselves had so efficiently been consuming for millions of years.

The winds of change have since repeatedly come and gone. The cranes of Europe ignored the human repression and Black Death of the Middle Ages, those of Asia have witnessed one horde of human invaders after another cross the central Asian steppes in vain dreams of glory, and the North American cranes have survived the plunder of a continent's natural resources in a minisecond of geological time. During the past century humans have managed to put nearly half of the world's cranes at risk, at the very time that we might do well to listen to their ethereal calls, which drift out over space and time in a haunting and somehow omniscient cry that carries both the authority of history and the urgency of reality. Cranes learned long ago of the need for social living in an indifferent or hostile world, of the value of prolonged and intense parental care, and of concern for the safety of the flock in the face of danger. They have seen mountain ranges rise and crumble, have watched entire civilizations rise and fall, and have observed great climatic changes that sometimes brought other animal groups to extinction. Yet each year they dance with an exuberance that gives joy to anyone with the eyes to see it, or even the imagination to visualize it. They seasonally cross entire continents with a precision that makes our best instruments seem inadequate, and fly with a breathtaking beauty that must make every pilot more than a little envious.

Humans have assigned themselves the unique privilege of determining which of the world's endangered plants and animals are worth saving and which are not. In making such decisions, we must be able to look beyond the obvious. Cranes will never allow themselves to be fully domesticated, nor will they ever provide humankind with a source of unlimited food or eggs. They are here in part to remind us that there should always exist a few wild places on earth where only very special animals can survive. Such animals carry with them unspoken messages from those remote and wonderful places that only they can visit easily. Most people have never seen the Himalayas, nor will most have the good fortune to seek out black-necked cranes on the Tibetan Plateau. Yet perhaps it is enough to know that the mountains are there, and that somewhere amidst those mountains there is a wonderful species of crane whose home is still largely a mystery, and whose life is still essentially untouched by human influence.

# THE
# SANDHILL
# CRANE

---

---

## A HISTORICAL
## INTRODUCTION
## TO THE
## SANDHILL
## CRANE

---

The sandhill crane first officially entered the world of ornithology in 1750 when George Edwards, in his *Natural History of Uncommon Birds*, illustrated a "Brown and Ash Colour'd Crane" from the vicinity of Hudson Bay, Canada. A few years later, Linnaeus classified it with the herons in giving it its first formal Latin name, *Ardea canadensis*. It was known to scientists by this name until 1819, when it was transposed into the crane genus *Grus*; and *Grus canadensis* is still the Latin name borne by the sandhill crane. However, for a long period this species was believed by many ornithologists, including no less a personage than John J. Audubon, to represent simply an immature plumage phase of the whooping crane.

In 1794, F. A. Meyer described a new variety of crane from Florida, which he believed to represent a new species and which he named *Grus pratensis*. The new species was based on a brief account of a "Savanna Crane" (the Latin name *pratensis* refers to its savanna-like habitat) by the American naturalist William Bartram, who had encountered the bird in Florida and reported that it "made excellent soup." The Florida sandhill crane, as this race is now known, currently ranges across much of peninsular Florida and extends north into southern Georgia in the vicinity of Okefenokee Swamp. It formerly also nested in extreme southern Louisiana, but has long been extirpated from that state.

Among the largest of the races of sandhill cranes by adult weight, the Florida sandhill consists of an entirely nonmigratory population. This population is especially associated with the original wet "prairies" of central and southern Florida, where stands of sawgrasses once grew over vast areas, interrupted by scattered saw-palmettos and cabbage palms and by clumped hammocks of pines and hardwoods on somewhat drier substrates. Perhaps 4,000 Florida sandhill cranes currently exist in that state. They are most numerous in the Kissimmee Prairie region of central Florida, although drainage and housing developments have increasingly infringed on their breeding habitats. On the other hand, the establishment of large areas of improved pasturelands and associated present-day grazing practices have provided increased foraging opportunities for Florida sandhills. Thus, although suitable breeding localities may locally be declining for these birds, their overall habitat conditions are somewhat improved over earlier times, and their numbers are probably increasing locally.

In 1905, Outram Bangs and W. R. Zappey reported that sandhill cranes breeding on the Isle of Pines, south of Cuba, are slightly darker than those from Florida, and the birds are also somewhat smaller and shorter. These authors designated this population as a new species *Grus nesiotes* (the name meaning "an islander"), but suggested that it should probably be regarded eventually as a new subspecies of *canadensis*, as indeed it now is. Even at the time of its discovery, this population was probably not very large, although it apparently occurred not only on the Isle of Pines but also widely distributed across Cuba. Probably the densest populations occurred on the Isle of Pines, and in 1990 that island was judged to still support a remnant population of about 20 birds. A similar number also still existed in Cuba's Zapata swamp area, and about a dozen survived in the vicinity of Pinar del Río. The Cuban sandhill crane is regarded now as a highly endangered race, and its total population in 1990 was thought to consist of only 54 individuals.

In 1925, James Peters concluded that sandhill cranes breeding in the western and northern United States were separable from those in Florida, as well as from the much smaller arctic-breeding cranes, and suggested the name *Grus canadensis tabida* for this race. The name *tabida* means "shrinking" or "wasting away," and presumably referred to its then-shrinking population and habitats. More recent observations have shown that at least four geographically separate populations fall within the limits of this race, whose vernacular name is the greater sandhill crane.

One of these is the eastern or Great Lakes population, which consisted in the mid-1980s of more than 16,000 birds nesting from Michigan and Wisconsin west through the northern half of Minnesota and into southwestern Ontario and southeastern Manitoba. In only two decades this population has responded to conservation efforts, increasing fourfold in numbers. Birds of this population that nest from eastern Minnesota to Michigan migrate southeastwardly to winter in Florida, with nearly all of the birds stopping for a time during fall migration at the Jasper-Pulaski Fish and Wildlife Area, in northern Indiana. Those from northwestern Minnesota and adjacent Canada fly almost directly south to winter along the coast of Texas.

A second population breeds in the Rocky Mountains, mostly

from Montana and eastern Idaho south through western Wyoming and into northern Colorado. This is the largest single population component of greater sandhills, consisting of 17,000–20,000 individuals in the mid-1980s. These birds winter mainly along the Rio Grande of New Mexico, with some occasionally reaching northern Mexico.

A third relatively small population of a few hundred cranes nests in northeastern Nevada and adjoining southern Idaho, and winters along the lower Colorado River. Finally, in southern and southeastern Oregon and adjoining northeastern California there is a "Central Valley" population of a thousand or more greater sandhills that winters not far to the south in the Central Valley of California. This is the only population of greater sandhills that is declining in at least some parts of its range, primarily as a result of high levels of nest and chick predation by common ravens, coyotes, and raccoons.

In 1966, an intermediate-sized "Canadian" race breeding between the ranges of the temperate grassland- and forest-breeding greater sandhill crane and the arctic tundra-breeding lesser sandhill crane was described by Lawrence Walkinshaw, and was called *Grus canadensis rowani*. Named after an illustrious Canadian ornithologist, William Rowan, this Canadian race of sandhill cranes is poorly studied. Its breeding grounds are scattered widely in boreal forest bogs and other subarctic wetland habitats across central and southern Canada from British Columbia to eastern Ontario. The birds from Manitoba and Ontario winter along the coast of Texas together with some greater sandhill cranes. However, little is known of the movements of the more westerly populations, which also winter with greater sandhill cranes of the Rocky Mountain and Pacific flocks. Because of the difficulties of separating Canadian sandhill cranes from greaters at the larger extreme, and from lessers at the smaller end of their size range, little can be said with certainty of the population size of this ill-defined, transitional race.

The recognition of the Canadian race of sandhill crane left

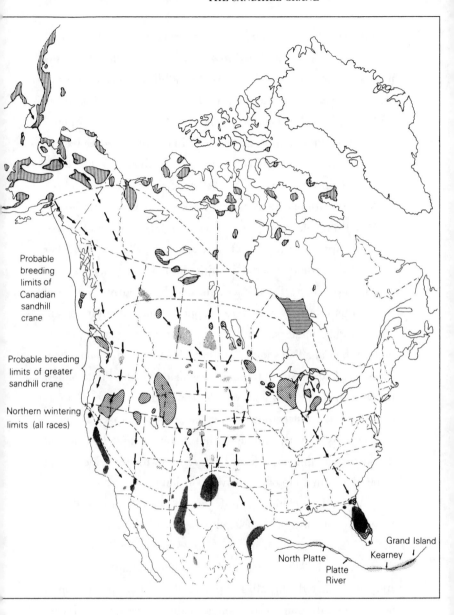

Breeding ranges of lesser (vertical hatching), Canadian (horizontal hatching), and greater (diagonal hatching) sandhill cranes, and residential ranges of Mississippi and Florida sandhill cranes (cross hatching). Lighter stippling indicates migratory staging areas, and denser stippling indicates major wintering areas of migratory races. The inset (lower right) shows the Platte Valley of Nebraska, with major spring crane concentration areas cross-hatched. Arrows indicate approximate fall migration routes. Modified from Johnsgard (1983).

35

only the small arctic tundra-breeding forms of sandhills as "less-ers." This subspecies (*Grus canadensis canadensis*) is easily the most abundant of all of the races of sandhill cranes, with population estimates over the past decade ranging from 200,000 to 500,000 birds. Besides the primary or "midcontinental" population of lesser sandhill cranes that winters in the southwestern Great Plains, a subsidiary population of about 25,000 birds winters mainly in the Central Valley of California and breeds along the southwestern coast of Alaska from Cook Inlet to the Alaska Peninsula and Bristol Bay. Careful visual and photographic surveys during March in Nebraska's Platte Valley offer the best opportunities for estimating the much larger, midcontinental population of this race, inasmuch as during that brief period this relatively small area supports more than 90 percent of the world's lesser sandhills (plus a few greaters and probably most of the Canadian sandhills). In 1985 a total spring population estimate of some 515,000 cranes was made for the Platte Valley, the vast majority of which were lesser sandhill cranes. This is one of the highest estimates ever made for the lesser sandhill, and perhaps is overly optimistic, but it does seem to be true that the lesser sandhill is at least holding its own, if not increasing in numbers, in spite of extensive sport hunting.

A small residential population of sandhill cranes that is native to southeastern Mississippi was described by John Aldrich in 1972 as a separate race, *Grus canadensis pulla*. First discovered nesting in Mississippi in 1938, when perhaps as many as 100 birds existed, the Mississippi sandhill crane population is now essentially limited to a single county (Jackson) in Mississippi, although it may once have also occurred in Louisiana and Alabama. By 1975 it was thought to consist of only 10–15 breeding pairs, and to comprise a total population of 30–50 individuals. At that time a special Mississippi Sandhill Crane National Wildlife Refuge was established, which ultimately grew to include more than 15,000 acres. This Mississippi population is still the most endangered of all the races of North American cranes. Its wet savanna habitat has been seriously affected by drainage and conversion to pine plantations. Housing developments and various highway construction activities have exacerbated the problems facing this tiny population, including the recent construction of a section of interstate highway and associated exchanges through the heart of its breeding range. Only 11 active breeding territories were known to exist in 1990, and breeding success has evidently been hampered as a result of reduced genetic diversity.

Although cranes have been studied by many ornithologists at various times, one person whose life has fallen completely under the spell of sandhill cranes is Lawrence Walkinshaw. A Michigan dentist and dedicated bird watcher, Walkinshaw first observed migrating sandhill cranes in 1921. In 1930 he encountered them on their Michigan breeding grounds and became fully captivated. He subsequently spent fifteen years studying sandhills from the Alaska tundras to the Isle of Pines, and traveled some 70,000 miles while preparing his monograph *The Sandhill Cranes* (1949). No sooner had he completed that book than he began preparations for a monograph of all the cranes of the world. Retiring from his dental practice in 1968, he devoted most of his time during the next few years to the writing of his *Cranes of the World*, which was based on observations in the wild of all the world's cranes except for the black-necked crane. This book was published in 1973 and, together with his earlier one on the sandhill cranes, still provides a basic source of information for all crane biologists.

## SEASONS OF THE SANDHILL CRANE

### A Sandhills Spring

There is a river in the heart of North America that annually gathers together the watery largess of melting Rocky Mountain snowfields and glaciers and spills wildly down the eastern slopes of Colorado and Wyoming. Reaching the plains, it quickly loses its momentum and begins to spread out and flow slowly across Nebraska from west to east. As it does so, it cuts a sinuous tracery through the native prairies that has been followed for millennia by both men and animals. The river is the Platte.

There is a season in the heart of North America that is an unpredictable day-to-day battle between bitter winds carrying dense curtains of snow out of Canada and the high plains, turning the prairies into ice sculptures, and contrasting southern breezes that equally rapidly thaw out the native tall grasses and caress them gently. The season is sweetened each dawn by the compelling music of western meadowlarks, northern cardinals, and greater prairie-chickens, and the sky is neatly punctuated throughout the day with skeins of migrating waterfowl. The season is spring.

There is a bird in the heart of North America that is perhaps even older than the river, and far more wary than the waterfowl or prairie-chickens. It is as gray as the clouds of winter, as softly beautiful and graceful as the flower heads of Indian grass and big bluestem, and its penetrating bugle-like notes are as distinctive and memorable as the barking of a coyote or the song of a western meadowlark. The bird is the sandhill crane.

There is a magical time that occurs each year in the heart of North America, when the river and the season and the bird all come into brief conjunction. The cranes begin to arrive in Nebraska's Platte Valley about the end of February as the Platte begins to become ice-free. They funnel into the valley from wintering areas as far away as northern Mexico, but primarily from eastern New Mexico and adjoining Texas, where a variety of shallow, alkaline lakes have offered them safety through the coldest months. These areas are all at least 600 miles from the Platte, the equivalent of a 12-hour nonstop flight at 50 miles per hour. Some of the birds do stop en route, but probably the majority make the flight in a single day. They achieve their maximum air speed with the aid of south winds, and fly in uniformly spaced gooselike formations for optimum flight efficiency. As they reach the Platte near sunset, the formations begin to break up, and the birds start to circle above the river, looking for safe nighttime roosting sites. The occasional calls of the migrating birds gradually build into a deafening crescendo of crane music. Individual flock members try to maintain voice contact with parents, mates, and offspring as they begin to pour into roost sites on the river, and the darkening sky becomes a maelstrom of circling and descending birds.

Over 90 percent of the sandhill cranes using the Platte Valley in spring are lesser sandhill cranes, the smallest of all the races of sandhills and the one with the longest annual migration, from the American southwest to the arctic tundras of North America and eastern Siberia. At the time of their arrival in Nebraska the birds weigh about six and a half pounds, and they stand about four feet tall. Like all other sandhill crane races they are grayish in plumage, but the crown of birds at least a year old is bare and the skin is bright red.

A small percentage of the sandhill cranes on the Platte are larger-sized birds, averaging perhaps eight and a half pounds and with proportionately longer bills. These birds, Canadian sandhill cranes, are headed toward subarctic nesting areas in Ontario and the other interior provinces, to muskeg or boggy openings in the vast coniferous forest that covers the heart of Canada. A very few represent greater sandhill cranes, the largest of the migratory races of sandhills. These large and distinctively long-billed cranes often weigh ten pounds or more. Those using the Platte Valley are headed for nesting areas in northwestern Minnesota, but most of these very large sandhill cranes have quite different migratory routes that pass either well to the west or east of Nebraska.

Counting all races, the sandhill cranes in the Platte Valley build up to a total of perhaps 400,000–500,000 birds by late March. This number includes essentially all of the lesser sandhill cranes occurring east of the Rocky Mountains and is not only the largest concentration of sandhill cranes in North America, but easily the largest crane concentration in the world. (The other extremely widespread and next most common crane species, the Eurasian crane, has an overall world population of about 100,000 birds. Its largest reported migratory and wintering concentrations number only about 20,000 individuals.) The Platte Valley and the adjoining shallow marshes of the "Rainwater Basin" immediately to the south also host about a quarter-million greater white-fronted geese, or most of those that migrate through the interior of North America. Vast numbers of Canada geese, snow geese, and wild ducks, especially northern pintails and mallards, also migrate through the area. The overall migratory waterfowl numbers annually total about seven to nine million birds, one of the most spectacular concentrations of migratory birds to be found anywhere in the world.

Additionally, up to a hundred or more bald eagles often winter on the Platte. While the eagles normally feed mostly on dead or dying fish, they occasionally fly over and harass the flocks of ducks and geese, apparently to determine if any crippled or partially disabled birds might be present and perhaps provide fairly easy prey. They pay little attention to the sandhill cranes, whose sharp beaks are likely to pose a serious threat to an eagle, and rarely does the sight of an eagle put a crane flock to flight.

The origins of the long-term love affair between the sandhill cranes and the Platte River are lost in prehistory. The oldest known evidence suggesting its antiquity is a fossil humerus, or upper arm bone, found in Miocene deposits of western Nebraska that date from about nine million years ago. It has a structure virtually identical to that of modern sandhill cranes and, if accurately identified, represents not only the oldest sandhill crane fossil ever discovered, but also the oldest fossil attributable to any modern species of bird. At that time in Nebraska's preglacial history, the landscape was evidently a grassland somewhat similar to today's, but having an associated mammalian fauna more like that of present-day East Africa than of North America, with rhinos and horses instead of bison and domestic cattle.

More definite evidence of long-term sandhill crane use of the Platte comes from the writings of various early explorers such as John Thompson, who in the spring of 1834 reported seeing sandhill cranes gathered on the Platte. A somewhat later account was provided by a hunter-adventurer who described his attempts to stalk a large flock of sandhill cranes near Grand Island during the fall of 1841. These were among the earliest explorers and immigrants to use the Platte as a convenient overland route leading into the western wilderness. By the mid-1800s the Platte Valley of Nebraska Territory was to become the primary route leading to Utah and the Oregon Territory. During that time tens of thousands of people followed the Mormon and Oregon trails beside the Platte on their way to new lives and fresh frontiers. Doubtless the cranes and waterfowl of the Platte provided important sources of food along the way.

At this time, even though the Platte was generally placid, it was still a surprisingly treacherous river for much of its length, being both "too thick to drink and too thin to plough." Its generally shallow, muddy, and wide channels could easily hide quicksand-like bottoms, and its annual spring floods could easily carry away both men and their horses or livestock. Its innumerable channels were constantly adding to and subtracting from the land, producing new sandbars and islands as rapidly as other ones were erased. Its banks were kept almost wholly free of trees by the spring floods and ice floes, and especially by the lightning-set fires that periodically raged over the prairies.

It is hard to know just what the attraction of the Platte River was to sandhill and whooping cranes in presettlement days, but probably its wide channels and vegetation-free islands provided ideal protection from prairie wolves and coyotes, while the adjacent wet meadows certainly offered protein-rich foods in the form of seeds and invertebrates. In the century and a half since the first white explorers described these flocks, the river has changed greatly. Most obviously, about three-fourths of its volume has been lost as irrigation projects have diverted its flows. The once-raging spring floods that carried mountain meltwater down to the Missouri River have largely been replaced by dried or cut-off channels. Its once grassy or shrubby shorelines, now protected from uncontrolled prairie fires, have grown up to gallery forests lining the riverbanks. Finally, its innumerable islands have become shrub- and tree-covered as the annual ice-scouring effects of early spring flooding have been progressively diminished.

With the loss of many of the Platte's historic channels, there has been an ever-increasing crowding of the birds into the few remaining acceptable roosting sites. These sites are now limited to a stretch of less than 100 miles of river distance between Kearney and Grand Island along the middle reaches of the Platte in east-central Nebraska. As a result, a population that was once distributed along at least 200 miles of river is now concentrated into fewer than 20 major roost sites, most of which are not on protected land and are subject to varying degrees of human disturbance.

Two major wildlife sanctuaries have recently been established on that part of the river that offers the best remaining crane habitat—the Audubon Society's Lillian Annette Rowe Sanctuary near Kearney, and the Mormon Island Crane Meadows sanctuary of the Whooping Crane Trust, near Grand Island. Both offer riverside blinds from which people can watch the daily drama in relative comfort and, more importantly, without unduly disturbing the cranes. The Rowe Sanctuary was funded by a single bequest, while the Crane Meadows sanctuary came about as the result of an environmental settlement in federal court. This settlement established a fund of more than seven million dollars, to be used for mitigating critical habitat losses for whooping cranes caused by the building of Grayrocks Dam on the Laramie River in Wyoming (a tributary of the North Platte, the single major source of water for the Platte). The other primary source, the South Platte, has already been seriously dewatered. Thus, in spite of the existence of these two important sanctuaries, the historic ties between the cranes and the Platte River are not guaranteed in perpetuity, and the conflicting needs of wildlife and the potential human exploitation of the Platte's water are likely to be brought into ever sharper focus in the future.

If the Platte has become so seriously degraded in recent decades, what then is it that draws the cranes back to it each year? The Platte still offers nighttime protection in the form of scattered sandbars and islands, though in ever fewer sites. Perhaps more importantly, the once vast wet meadows have largely been replaced by cornfields, in which the birds can feed daily for as long as five or six weeks, eating unharvested corn left over from the previous summer. In this way they can quickly build up their fat reserves to a maximum, adding about a pound of fat to their total body weight and putting them into ideal condition for their long remaining journey to the arctic. They must arrive on the nesting grounds in prime physiological readiness to breed, for there will be very little to eat during the first few weeks on the tundra.

Each day while the cranes are in the Platte Valley they leave their river roosts shortly after sunrise, as pairs, families, and small flocks spread out both north and south from the river to forage in nearby cornfields. They also feed in the few remaining wet meadows, where invertebrates still provide their best sources of high-protein foods. Each evening they return to traditional roost sites, each of which holds about 10,000–15,000 birds, located in the least disturbed portions of the river well away from bridges and easy human access. To these same roosts vast numbers of cranes return every night near sunset after they have finished their daytime foraging activities. The sunrise and sunset flights of tens of thousands of cranes provide a sight that overwhelms the senses, the din of the birds almost making one dizzy, and the sight of the wheeling flocks overhead seeming at times like a scene from fantasy or science fiction.

At almost any time while on their roosts or while foraging,

"dancing" behavior may suddenly begin. This consists of bows, jumps, vegetation-tossing, and wing-flapping activities that are not limited by age or sex. Dancing may quickly spread through a small group of birds, and may just as suddenly end. Sometimes it is started by a sudden, possibly frightening stimulus, and under such circumstances the bounding movements of the birds may quickly change to actual flight, but at other times no apparent stimulus is evident. Although crane dancing vaguely resembles some primitive forms of human dancing and as such has been traditionally believed to represent courtship, in fact it probably has relatively little to do with pair bonding.

Cranes pair for life, and so true courtship is needed only infrequently. Instead, various pair-bonding activities by adults, such as "unison calling," serve periodically to reinforce existing pair bonds. In sandhill cranes, unison calling is done simultaneously by both members of a pair, the female usually uttering about two calls per male call and not throwing her head so far backward during the call as does the male. Perhaps these sex differences during unison calling help to reinforce sexual identity, and thus help to avoid same-sex pairings.

By early April, many of the sandhill cranes have begun to leave the Platte Valley, often beginning their migration by gaining great altitude, wheeling about in massive flocks that rise slowly, their broad wings riding the thermal updrafts produced by the warming April sun. Even before they leave, the birds often spend hours in such circling flocks above the Platte Valley, perhaps simply reveling in the sheer joy of such low-energy flying, or perhaps using these high-altitude maneuvers as reconnaissance flights to scan the river and commit its topographic features to the collective memories of the flock members. This procedure may be especially important for the younger and more inexperienced birds, which must eventually learn all of the species's most secure migratory stopping points along their several-thousand-mile journey.

*A Teton Summer*

In some parts of the northern Rocky Mountain region, such as the vicinity of Grand Teton National Park, a thriving population of greater sandhill cranes nests each year. These birds are already establishing nesting territories by late April and early May, long before the lesser sandhills have reached their arctic breeding grounds. Here, along the Snake River, the birds share their habitats with another large and very rare species, the trumpeter swan, as well as with a wide variety of other wildlife, including beavers. Indeed, the swans and cranes are closely associated ecologically with beavers, for the dam-building activities of these animals provide small, stable impoundments that offer ideal nesting sites for both the swans and the cranes.

As soon as their nesting habitats become snow-free, the greater sandhills of the Rocky Mountain region begin to seek out suitable nesting territories. Ideally these must have not only excellent nesting locations, but also nearby foraging areas where the soil is soft enough for them to dig up spring plants for their edible roots, and where some high-protein foods are also available. The cranes are not averse to eating the eggs or chicks of smaller marsh-nesting birds that they happen to come upon, such as those of rails or red-winged blackbirds, and they may even eat the eggs of ducks as large as teal. There are a limited number of ideal wetland territories in this essentially semiarid environment, except where beaver activity has provided an abundance of ponds. Pairs sometimes fight fiercely for ownership of prime territories. Probably most of these fights involve the males of the respective pairs, although their mates take great interest in the contests, occasionally trumpeting encouragements and perhaps getting in an occasional peck or two. The fights may even attract others, setting off a general frenzy of activity.

Fights by cranes are similar to normal crane "dancing," which provides evidence for the view that such dances represent a variably ritualized or stereotyped version of aggressive behavior, including a jumping and simultaneous kicking, an occasional ground-pecking movement or tossing up of vegetation, and bowing or head-tilting movements that expose to its opponent the brilliantly red crown skin of the displaying bird. In fact the crown skin represents an excellent and conspicuous external clue to the internal state of the bird; relaxed sandhill cranes can retract the skin so that it barely extends back to the eye, but extremely excited birds can pull it back to the rear of the crown as it becomes engorged with blood. The aggression of cranes can also be directed toward other animals that perhaps represent a threat to their eggs or young, such as skunks. At times they have even been observed to stand up to such large animals as deer or moose when these have happened to wander too close to their nest.

Sandhill cranes do their best to hide both their nests and themselves during the breeding season. They achieve this in part by their inconspicuous gray adult plumages, and even this color is effectively improved in its camouflage value by the birds staining their body feathers with mud and rotting vegetation prior to the start of nesting. This "painting behavior" has only rarely been observed in wild individuals, inasmuch as the birds become extremely secretive just prior to nesting. Evidently it is done at odd times when the birds are not otherwise engaged in foraging or other activities. The birds dredge such materials from pond bottoms with their bills and spread it over their plumage, causing the feathers to become variably brownish from the soil and organic pigments. Gradually their body feathers become much the same color as the dead vegetation of their nesting environment, although the head and neck, which the bird cannot reach, remain unstained.

Just as mates are retained year after year for so long as both pair members remain alive, nesting territories are also reoccu-pied on a yearly basis by established pairs. Based on a sample of 45 known-age greater and Florida sandhills, it has been found that initial breeding attempts most often occur at three years of age. However, these first nesting efforts are usually failures, and some birds may not even attempt to breed until they are five years of age. Pair bonding usually begins during the latter part of the crane's second year of life. Because of ephemeral associations, the subadult crane is, on average, associated with five different potential mates before breeding successfully. However, after pair bonds are firmly established a pair might remain together for ten years or more, depending on the survival of both its members. In the studied sample, remating occurred fairly rapidly after the death of a mate, especially among males, which required from 5 to about 77 days to remate in four instances. However, one female required 132 days to remate, and another had still not remated 271 days after her mate's disappearance. In this same study it was found that almost half of the sandhill cranes (mostly of the Florida race) that were banded during or after their third year of life had been associated with more than one mate during varied periods of observation (ranging up to as much as eleven years).

Nest construction may require a week or more, with most of the work done by the female. Like trumpeter swans, sandhills slowly accumulate material on the nest as the bird stands or sits on it, reaches out, and pulls or throws material to the site. This pile is not lined in the least with feathers. Instead, the eggs are simply laid directly in the shallow nest bowl. Sandhill and whoop-ing cranes almost invariably lay two eggs, which are deposited at intervals of about two days. Incubation of the first egg begins immediately after it is laid. This is in marked contrast to swans and geese, which do not begin incubation until their large clutches are complete. This is an important distinction, for in species that begin incubation immediately, the eggs will hatch in the same sequence in which they were laid, and at roughly comparable intervals. However, if the female delays starting her incubation until the clutch is complete, all of the eggs can hatch almost simultaneously, and the brood can all leave the nest at the same time. This simultaneous hatching is probably especially advantageous for those bird species, such as geese and ducks, that usually have large clutch sizes and precocial young, but it may not matter so much for cranes, having only two eggs and thus no more than two offspring to look after.

Fertilization occurs in the days prior to the laying of the first egg, mostly during the immediately previous week. During her egg-laying period, the female usually does not leave the nest at all, but neither is the male likely to visit it. However, as soon as the second egg has been laid the male is likely to approach the nest and gently ease his mate off it, so that he can begin incubating and she can go off to forage. From that point on, the pair takes turns incubating, with each parent staying on for several hours while the other forages or stands guard, watching for possible danger. Generally the female will take over in late afternoon and remain on the nest all night, with the male roosting nearby.

For 30 long days of incubation, the life of the crane pair is thus centered on the nest. Meantime, other signs of spring and early summer are everywhere to be seen or heard. The dawn drumroll of a displaying ruffed grouse softly penetrates the stillness of the dense evergreen woods, scarlet gilia flowers begin to blossom at the edges of woodlands, attracting bees and especially calliope hummingbirds, and in more open sunlit areas early summer flowers begin to give dashes of color to the grayish green sagebrush-dominated upland flats. Here and there a mule deer doe delicately moves through the woods, and cow moose lead their newborn calves through beaver ponds, searching out beds of water lilies, one of their favorite foods. Cow elk have moved up from the open woodland and sagebrush flats where they spent the winter into cool mountain meadows, where they each give birth to their single young calves in isolation. Within a few weeks of their births, the young calves are brought together by their mothers. There they form a calf "pool" that can be collectively guarded by one or two of the females, while the others can take time to feed and generally recuperate from the stresses of calving.

During the hatching period the female crane sits tight on the nest, while the male stands close guard nearby. The hatching period of sandhill cranes is a critical time, as it is for all birds. As much as 24 hours may elapse from the time of initial eggshell pipping by the embryo until the baby bird finally kicks free of the shell, which is done without any assistance from the parent bird. During this time the female sits even more closely than before, only occasionally standing up to look at the hatching egg or sometimes move it gently with her bill tip. After breaking free of the shell, the chick will lie for a time wet and helpless in the nest, resting from its ordeal and waiting for its wet down to dry out. During that time it slowly transforms from an almost shape-less mass into a beautiful creature covered with golden-coppery down, with a paler buffy ring around the eye. The fluid that initially swells the legs and toes becomes absorbed into the body, and within a few hours after escaping its shell the chick may begin to try standing on its feet. It huddles under the female's breast feathers for a time, but soon is peering out at the world, or perhaps even trying to climb up on its mother's back.

Shortly after each chick has hatched, the parent bird picks up the empty shell and associated membrane and either discards them near the nest or, more commonly, eats them. She may also break the shell into tiny fragments and hold them up in front of the chick's bill, stimulating it to nibble them.

Because of the approximately two-day difference in the ages of the chicks, the first-hatched chick is already adept at walking and swimming while the younger one is still being continuously brooded by their mother. This age difference, although seemingly minor, is actually of great importance in crane breeding biology. The competitive aggressiveness of the birds is exhibited even at this early age, and, unless prevented by the adult, the older chick is very prone to peck at its younger sibling. The intensity of such sibling aggression usually increases rather than diminishes as the chicks become older, and may even lead to the death of the weaker chick, usually the younger one. This aggressive behavior among the chicks is a rather puzzling aspect of crane biology, since natural selection should favor behavioral patterns that maximize the survival of the young. The young do not strongly compete for food, at least after the first few days when they are being fed almost entirely by their parents and pick up few if any food items on their own. Nevertheless, this sibling aggression might help account for the small, two-egg clutch size that is typical of nearly all cranes, inasmuch as larger numbers of chicks would probably tend to increase aggression to an even greater degree than already exists.

As the first-hatched chick begins to gain strength and starts to wander about increasingly far from the nest, the male parent usually steps in to take over its care. This arrangement, physically separating the two chicks from one another, greatly reduces opportunities for fighting between them, and each parent takes on the responsibilities for caring for a single offspring. Each chick closely follows behind one of the parents, while the latter searches out choice food items, such as insects. When food is

found, the parent picks the item up and holds it at the chick's eye level, waiting for the youngster to take the morsel. Soon the chick is finding prey on its own.

When the second chick is able to leave the nest, the female encourages it to follow her, and shortly thereafter the family has completely abandoned the nest site. The families tend to move gradually into heavier cover, where they can rarely be observed. Here they will spend the approximately two-month period required for the young to gain the power of flight. During that time the weight of the young will increase from the few ounces at hatching to nearly six pounds, and they will stand nearly as tall as their parents. However, their high-pitched "baby voices," cinnamon-tinted juvenal feathers, and fully feathered crowns allow for ready recognition of juveniles through most of their first year of life.

### North to the Arctic

As the cranes of the Rocky Mountain region are busy with their nests in May, those that left their spring staging areas in the Great Plains in April have begun the last leg of their spring migration across the interior of Alaska. Their tundra breeding grounds are still snow-covered and icebound through all of May, and thus the birds must wait in subarctic areas for the breeding grounds to become accessible. As they make their way through the interior of Alaska in early to mid-May, the cranes follow the northern edge of the Alaska Range. As rapidly as the weather allows, they move westwardly along the Tanana River valley and on into the lower Yukon-Kuskokwim drainage and the associated lowland tundra breeding areas along the Bering Sea.

Other lesser sandhills from Pacific coast wintering grounds are simultaneously passing north along the southeastern Alaska coast. These birds stage briefly along such traditional locations as the Stikine River delta, the Gustavus flats near Glacier Bay, and the eastern Copper River delta. These flocks are headed for breeding areas around Cook Inlet, along the north coast of the Alaska Peninsula, or on the lowland tundras of adjoining Bristol Bay.

The cranes migrating through interior Alaska often roost on river overflow ice of the Tanana River, or on the ice of ponds and lakes, as they wait for the tundra of western Alaska to become accessible to them. Along the much milder and snow-free Copper

River delta on Alaska's southern coast, the cranes reach their peak numbers in the first week of May, with most of them flying over the delta rather than stopping there. They thus overtake or bypass vast flocks of migrating shorebirds that are also headed for arctic breeding areas.

These tidal flats of the Copper River, enormously rich in invertebrate life, serve as a final staging area for uncounted millions of arctic-nesting shorebirds, a role comparable to that of the Platte River for lesser sandhill cranes. Shorebird flocks numbering in the tens or hundreds of thousands frequently build up in this delta in mid-May, when there may be as many as a quarter-million shorebirds feeding in a single square mile of tidal flats. Shortly after mid-May the shorebirds make a rapid and spectacular departure, headed in a northwesterly direction for breeding grounds in western Alaska and probably also northeastern Siberia.

Similarly, some of the lesser sandhill cranes migrating through interior Alaska will continue on across the Bering Sea toward arctic breeding grounds in northeastern Siberia. How many cranes actually do so is uncertain, but Soviet ornithologists have recently estimated the USSR population as at least 25,000 birds and possibly more than 50,000. Evidently the Soviet population is still increasing, and its range expanding. At least some nesting may now occur almost as far west as the Khatanga River, nearly 2,000 miles west of the Bering Sea. Banding data indicate that cranes nesting in eastern Siberia migrate across the Bering Strait and winter in Texas and New Mexico at least 4,000 miles away, possibly nearly 6,000 miles for birds nesting as far west as Siberia's Khatanga River. This is the longest known migration of any crane species, and perhaps is unmatched by any nonpelagic bird of comparable size. If, instead, the birds simply flew south to central China, as do the Siberian cranes that nest in the same general region, the sandhills could reduce their total migration route by about half. They would thereby also avoid the dangerous

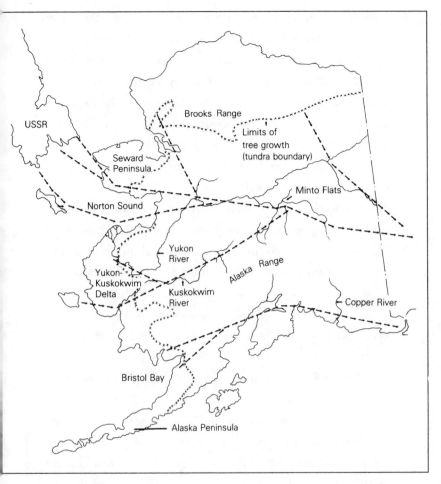

Fall migration routes of the lesser sandhill crane in Alaska. Spring routes are essentially the reverse of those shown for fall. Adapted with modifications from Kessel (1984).

Bering Strait crossing, which is frequently stormy and often fog-bound for much of the year.

By late May, the lesser sandhills are on the final leg of their spring journey, breaking out beyond the limit of trees in western Alaska and into the trackless wilderness of lowland coastal tundra that centers on the deltas of the Yukon and Kuskokwim rivers. This is the greatest breeding ground for waterfowl, cranes, and shorebirds in all of North America, and about a quarter-million cranes, or perhaps half of the world's total sandhill crane population, will scatter out to breed in the lowland tundras of western Alaska and adjacent Siberia. It is an environment having almost as much water as land, with an endless maze of lakes, streams, ponds, and marshes present during the brief summer. This complex interplay of water and permafrost-bound land provides a variety of seasonal wetland habitats ranging in size from tiny ponds to large lakes as well as coastal beaches.

The lesser sandhill crane is one of the most conspicuous species nesting on the vast tundra flats of the Yukon-Kuskokwim delta. Here at least two-thirds of Alaska's total breeding sandhill crane population spend the brief summer to nest and raise their young. Lesser sandhills, like the greater sandhills farther south, are highly territorial, with their nests widely scattered (averaging considerably less than a nest per square mile) and usually well hidden.

The problems of nesting in arctic tundra are somewhat different from those faced by the cranes nesting in the Rocky Mountains, with the much shorter available breeding season being the most important factor. Snow or freezing rain is possible almost any time, and even during the warmest months freezing rain may mercilessly pelt the land, threatening to chill adults, their eggs, or their young. Perhaps because of this, newly hatched lesser sandhills have considerably more luxuriant downy coats than do the sandhill cranes breeding in the southern United States. Much under-the-wing brooding of newly hatched chicks is also typical of arctic-nesting sandhills. On one occasion a parent has been observed carrying a day-old chick on its back while standing and foraging, the only time such behavior has ever been observed in cranes. In the Banks Island population of lesser sandhills, the cranes have been observed to forage regularly on lemmings, using a special head-lowered peering posture that enables the bird to look down lemming burrows for its prey. At times the lemmings are captured in their burrows or, if necessary, chased about above ground until captured. Then they are shaken and stabbed to death, and finally torn to pieces.

Lesser sandhill cranes have also apparently been able to adapt to the much shorter breeding season as a result of their considerably smaller overall body size. Although this has not influenced the length of their incubation period, the newly hatched young require considerably less time to fledge than do the larger cranes breeding farther south. This shortened fledging period may be a combined result of their lighter body weights at fledging and the nearly 24-hour daylight conditions that permit them to forage almost continuously from the time they hatch until they are ready to begin their fall migration. The young fledge when they are only about two months old, rather than at about three months as in greater sandhill cranes, and they are thus vulnerable to jaegers, gulls, arctic and red foxes, and other predators for significantly shorter periods. They fledge by the latter part of August and begin to leave their Alaskan breeding areas in early September, scarcely three months after the spring arrival of their parents.

### An Autumnal Hegira

By the first of September the day lengths are already noticeably diminishing, and snow begins to accumulate on the protected northern slopes of the low Kuskokwim and Askinuk Mountains that rise above the tundra. Shorebirds and waterfowl have already left the Yukon-Kuskokwim delta. The cranes soon begin to leave, heading east across the Kuskokwim Mountains and up the Kuskokwim River, where they will join the equally large flocks of cranes coming from Alaska's Norton Sound and those that nested in northeastern Siberia.

As the birds move east along the north side of the Alaska Range, they often enter Denali National Park and move past the brooding and majestic presence of Mount McKinley, occasionally then flying at a height of 20,000 feet or more above sea level. Typically the birds migrate at a height of about 1,000 feet above ground level during the fall period. This altitude is somewhat lower than is typical during spring migration, when flight altitudes of 3,000 to 5,000 feet above ground are common. Perhaps one reason for this seasonal difference is that recently fledged young birds may be less able to reach such great heights very easily. During spring migration some crane flocks fly as high as about 7,500 feet above ground, apparently to avoid the sometimes severe air turbulence occurring closer to ground, which is more prevalent during spring than fall. Many flocks stop and spend a few days resting and feeding along the McKinley River bars, a habitat that is much like the Platte River. Even more will rest for a week or so on the Minto Flats, along the lower Tanana River. At times the cranes may have to fly in fog or snow when they remain close to the ground. At other times strong headwinds, poor visibility, or heavy rains halt their migration.

As the cranes pass through interior Alaska they pass scattered groups of caribou that are also engaged in migrations, the animals gradually moving back south out of their tundra calving grounds into heavier winter cover. These animals are soon to face the rigors of the fall rutting season and the even greater stresses of an Alaskan winter. In somewhat heavier cover than that used by the caribou and cranes, moose are also putting on fat for the winter, although they do not have the additional pressures of migration to contend with. Likewise, grizzly bears spend all of their daylight hours in their search for food. They are able to consume almost any sort of plant or animal food they encounter, including even carrion, but seem to relish the abundant fall crop of berries that give small sparks of cold color to the Alaska landscape during this period.

Most crane migration is done during daylight hours, with the birds stopping to rest each night, although in fine weather flights may continue well past sunset, especially on moonlit nights. The daylight period is shorter during the fall migration through Alaska than during spring, and thus the birds are more likely to encounter darkness on the longer legs of their journey.

Eventually the birds break out of the Canadian mountains in northern British Columbia and pass into the western Great Plains. From there they make a nearly straight-line southeastern flight through the Peace River area of central Alberta, and from there on into western Saskatchewan. Here they arrive in mid-September as the grain fields are being harvested and the native prairies and marshes are rich in natural foods.

As they arrive in Saskatchewan, the lesser sandhills begin to encounter small family groups of whooping cranes, which have also begun their fall migration out of the Wood Buffalo Park nesting area about 600 miles to the northwest. Many of the whooping crane pairs are now leading single offspring, already almost as large as their parents, but distinctly rust-colored on their head and upperparts. Also using this same region are great flocks of lesser snow geese and other arctic-breeding geese, all of which are hurrying southward to escape winter. Many of these birds will turn southwest and head for wintering areas in California, while some will go south to the Rio Grande Valley of New Mexico. Still others will follow a route very similar to the lesser sandhills', moving southeast into the Dakotas and then down the central and southern Great Plains into wintering areas of interior Texas and along the Gulf Coast.

Although the lesser sandhills and whoopers use this area simultaneously, there is little or no direct competition between them, as the whoopers tend to use wetland areas while the sandhills are feeding in natural prairies and grainfields. Both, however, must accumulate the important fat energy reserves to be used in making the last major leg of their fall migration, just as was the case during their stopover in the Platte Valley on the way north in spring. During this early autumn period the birds are likely to start encountering crane hunters, inasmuch as sandhill crane hunting has been legal in Canada since 1959 and in the United States since 1961. Estimated annual legal "harvests" by U.S. and Canadian hunters during the late 1970s and early 1980s have averaged about 14,000 birds, to which another 3,000 or so can be added to account for crippled birds that were never retrieved as well as birds shot by Mexican or Alaskan hunters and never reported. Fortunately, relatively few hunters have the patience, knowledge, and experience to hunt cranes very effectively. Additionally, most of the cranes that are killed are inexperienced and relatively unwary young of the year, the loss of which is not so serious as the destruction of paired and reproductively active adults.

Many of the effects of such "sport" hunting on the population ecology of lesser sandhill cranes are unfortunately still only poorly understood, such as the time required for the establishment of new pair bonds by widowed birds, or the effect of the loss of one or both parents on the chances for survival of their juvenile offspring. Although it is known that, like all cranes, lesser sandhills mate for life, it is not yet clear just when the cranes of this race normally mate and initially reproduce in the wild, or what their normal longevity might be under both hunted and protected conditions. As a result, population management of lesser sandhills has been as much a matter of guesswork as of informed scientific judgment. It is perhaps a testimony to the innate wariness of the birds and their adaptability to changing conditions that the North American lesser sandhill crane has apparently not yet been seriously impacted by the advent of legalized hunting, and these annual hunting-related losses probably so far account for no more than about 5 percent of its total population. Certainly some additional mortality unrelated to hunting must occur, such as fatal diseases and accidental deaths. Yet apparently all of these hunting and nonhunting losses collectively still represent less than the lesser sandhill's average annual recruitment rates of around 7–11 percent, and the population is certainly maintaining itself if not actually increasing. It is, however, surprising to most non-North Americans to learn that cranes, strongly symbolic of longevity and good fortune throughout much of the Orient and fully protected in almost all civilized countries, are treated in North America as fit targets for "sportsmen."

After spending several weeks foraging on the grain-rich Canadian prairies, the sandhills begin to leave in early October. Some of them turn to go southward through eastern Montana and follow the eastern foothills of the Rockies to wintering areas in southern New Mexico and adjacent Mexico, but most will continue southeastwardly through central North Dakota, stopping again for a time in the Dakota grainfields and shallow marshes. From that point on they have a nearly direct flight southward, crossing the Platte Valley of Nebraska but with few of the birds stopping there. Instead, their destination is the arid grasslands of eastern New Mexico and northwestern Texas, where they will spend the entire winter. The whoopers follow a similar route, but rather than going to western Texas they head directly for its Gulf coast, to the safety of Aransas National Wildlife Refuge.

The greater sandhill cranes of the Rocky Mountains, whose young fledged in late August or September, also begin to funnel southward during October. The birds of the northern Rockies pass south along the western slope of the Colorado Rockies until they reach the drainage of the upper Rio Grande in southern Colorado. From there they pass straight south along the Rio Grande of New Mexico, heading for wintering areas in central and southern New Mexico. Although most of the Rio Grande's crane habitats have been lost, a few important suitable wintering sites remain, of which much the most important is Bosque del Apache National Wildlife Refuge south of Albuquerque.

This refuge is beautifully situated at the foot of the arid Magdalena Mountains, which shimmer in pastel tones in the clear desert light. The area has probably been a haven to wildlife since prehistoric times, with Indian ruins dating back to 1300 AD. A variety of shallow ponds and lakes provide perfect roosting sites for the cranes and also offer protection to snow geese, Canada geese, and other waterfowl. Additionally, cornfields planted for wildlife use allow both the cranes and the geese to spend their time in relative safety, although some goose and deer hunting is allowed.

The annual results of the whooping crane egg-transfer experiment become evident each fall at Bosque del Apache, as greater sandhill cranes from Idaho begin to arrive with their whooping crane adoptive offspring. By this time the young whoopers stand even taller than their "parents" and are already starting to acquire the white and black plumage pattern typical of older age classes of whooping cranes. They appear to be fully accepted by the sandhills in spite of their larger size and alien appearance. There can be no question that lesser snow geese have responded to effective management efforts, and upwards of 50,000 of these beautiful birds have been attracted to Bosque del Apache refuge in recent years. Many of the even smaller Ross' geese are present as well.

A few hundred miles to the east, in the dry Staked Plains of the New Mexico-Texas border country, lesser sandhill cranes are also arriving on their wintering grounds. In refuges such as Bitter Lake National Wildlife Refuge, New Mexico, and Muleshoe National Wildlife Refuge, Texas, the birds pour into the shallow, often alkaline playa lakes that are scattered over this desolate landscape.

The shallow Pecos River passes slowly southward in its narrow U-shaped valley, and shallow areas of seepage provide marshes that conform to the outlines of now-ancient channels of this river. These alkaline marshes, fed also from underground springs that give rise to a creek called Lost River, form the lifeblood of Bitter Lake Refuge. This isolated and starkly beau-

Greater sandhill
flushing teal from nest
to get eggs.

tiful refuge is rimmed on most sides by low mountains and is characterized by barren salt flats, gypsum sinks, and especially by numerous shallow ponds and impoundments that offer life-sustaining water to the plants and animals of the area.

Other than migratory geese and smaller waterfowl, the wintering birds of Bitter Lake and Muleshoe refuges are primarily the lesser sandhill cranes. As they arrive they are finishing an epic journey that has spanned an entire continent in only about two months, from arctic tundra to desert wilderness, one they have reenacted annually for millions of years. Their pattern for survival has been fixed in an endless seasonal repetition of tundra, mountains, and plains, and cycles of annual birth, death, and rebirth. In remote and safe places like Bitter Lake and Muleshoe refuges the birds will finally rest and recuperate during the short winter days. And thus they will be ready for another spring flight northward, when the arc of the sun again turns to give warmth and light to the frozen northern lands.

# THE
# WHOOPING
# CRANE

## A BRIEF
## HISTORY OF
## THE SPECIES

If any single bird species symbolizes the North American conservation movement of this century, and the closeness many wildlife species came to extinction, it is the whooping crane. Probably never very common, the whooping crane population numbered perhaps less than 2,000 at the time of European settlement, but its breeding range probably extended broadly across the grasslands and marshes of interior North America. The first known record of the species in the ornithological literature dates to 1722, when the English naturalist Mark Catesby visited South Carolina and obtained the skin of a whooping crane from an Indian. Catesby rightly considered it a previously undiscovered species and named it *Grus americana alba*. However, it was not until more than a century later that the United

States National Museum finally obtained a whooping crane skin for its own collection.

Following the Civil War and the opening of the West to settlement, the whooping crane was increasingly encountered, and its breeding and wintering habitats were progressively altered and finally destroyed. The last three decades of the nineteenth century were especially disastrous for the birds, for during that period not only were they killed by market hunters, but also egg collectors and taxidermists became aware of the great value of whooping crane eggs and skins to museums and other collectors. It has been suggested that perhaps as much as 90 percent of the entire population was destroyed during that relatively brief period, when the rich grasslands of the upper Great Plains of the northern states and prairie provinces were also being converted to farms. Thus, nesting in Illinois was eliminated by 1880, and during the next ten years the birds were lost as breeders in Minnesota and North Dakota. During the 1890s the birds were also eliminated from Iowa, which represented the last known breeding record for the United States. However, in adjacent southern Canada the birds persisted longer, with a nesting pair discovered as late as 1922 at Muddy Lake, Saskatchewan. The nestling chick of this pair was collected for a Canadian museum, and thus ended the last known nesting site of the species in North America.

During this period it was known that a population of whooping cranes still existed in the coastal marshes of southern Louisiana. These birds certainly represented wintering birds from the Great Plains or farther north, and apparently also included a resident population of unknown size. During the late 1920s the U.S. Army Corps of Engineers extended the Intracoastal Waterway to Grand Lake, Louisiana, thus making accessible the vast areas of coastal marshes to hunters and the adjoining tallgrass prairies to farmers, who quickly set about converting the area to rice culture. By 1940 the Louisiana whooping crane population had declined to only 6 individuals, less than half of the previous year's total.

During the 1930s it became apparent that the major remaining whooping crane population consisted of birds wintering in coastal Texas in the Blackjack Peninsula area (between Aransas and San Antonio bays) of Aransas County. Their breeding grounds were judged to be somewhere north of the U.S. boundaries, most probably in the vast and largely unexplored regions of northern Canada. A critical moment in the history of the whooping crane's survival occurred during the late 1930s when the Bureau of Biological Survey (later to become the U.S. Fish and Wildlife Service) purchased about 75 square miles of the Blackjack Peninsula for habitat preservation. In December of 1937 it established the Aransas Migratory Waterfowl Refuge (later called the Aransas National Wildlife Refuge), and the wintering grounds of the whooping crane finally came under full protection.

This protection came not a moment too soon, for by the fall of 1938 only 14 adults and 4 juvenile cranes arrived on the newly created refuge to spend the winter. These, together with the 11 birds known to occur in Louisiana, represented a grand total of only 29 whooping cranes in existence. In spite of this protection the Texas population did not immediately respond, and perhaps an all-time low was reached in 1945, when only two birds were known to be surviving in Louisiana and 17 were found on the Aransas wintering area. Considering that some of these were juveniles, subadults, or unpaired and nonbreeding adults, the actual number of remaining breeding pairs may have numbered three or four! During the 1940s the Louisiana population declined from 6 birds in 1940 to only a single individual in 1948.

During this critical period, as the U.S. was finally emerging from the dark years of World War II, it became increasingly apparent that, if the species were to be saved at all, it would be necessary to locate and protect its breeding grounds. In 1945 the U.S. Fish and Wildlife Service and the National Audubon Society thus established a whooping crane project, agreeing to a joint sponsorship of field studies, research on crane biology, and especially a hunt for the whooping crane's breeding grounds. This almost impossible task was initially taken up by Fred Bard, Jr., of the provincial museum in Regina, Saskatchewan. Bard used historical migration and breeding records to narrow his field of search to various parts of eastern Alberta and Saskatchewan. His field work began in the spring of 1945, when he and Robert H. Smith, a Fish and Wildlife Service biologist assigned to the project, hunted fruitlessly through many of these areas. During the next breeding season the search was taken up by Dr. O. S. Pettingill, Jr., who similarly searched out potential breeding areas in Manitoba, Alberta, and Saskatchewan, partly with the assistance of Robert Smith, but again to no avail.

When Dr. Pettingill had to return to his teaching responsibilities that fall, a biologist from the National Audubon Society was named to replace him. This dedicated and indefatigable man, Robert Porter Allen, had already done important work on roseate spoonbill conservation, and was a perfect choice for such a daunting task. During the 1946–47 winter, Allen studied winter territoriality in the whoopers at Aransas and also made important observations on foods and foraging behavior, habitats used by the cranes, and the like. The next spring he set off for Canada to look for possible nesting areas in Manitoba, Saskatchewan, and Alberta with the help of Robert Smith. Like the previous searches, this too proved unsuccessful, and it became increasingly evident that the birds must be nesting farther north than anyone had previously suspected. Interestingly, as early as 1864 two nests had been discovered in Mackenzie District of Northwest Territories, one near Fort Resolution and the other on the Salt River. Both of these had been found in aspen parkland habitat, which was judged by Allen to be the most probable nesting habitat of Canada. These two nests were among the first ever discovered for the species, and therefore the general area south of Great Slave Lake had been briefly scanned from the air by Allen and

Smith in 1947. This effort also failed, perhaps because bad weather had caused greatly reduced visibility in the most promising area.

In 1948 Allen decided that the vast wilderness to the north of Great Slave Lake must be searched. Yet, after extensive aerial surveys of coastal and interior areas, not a trace of the whoopers could be found that summer. Thus four years of searches had all led nowhere. In frustration Allen turned to writing up his data on the species, which was published by the National Audubon Society as a special research report in 1952.

In 1952 the first real clue to the nesting grounds emerged when Robert Smith found two whooping cranes during July in an area of wilderness near Great Slave Lake, southern Mackenzie District. However, a second flight over the same area in August revealed no cranes. Another possible sighting of a single bird was made the next year in the same general vicinity, and during the fall of that year a flock of eight migrating birds was seen along the Slave River, just south of Great Slave Lake. Thus, a pattern involving the Slave River as a migratory pathway for breeding birds, if not for nesting itself, seemed to be emerging. By then, wintering counts indicated that some two dozen whooping cranes existed, suggesting a very slight upturn in their total numbers, but with no real results evident from the winter protection afforded by Aransas Refuge.

The critical break was finally made in 1954 by William Fuller, a Canadian Wildlife Service mammalogist stationed at Fort Smith on the Slave River. Fuller was doing a survey of mammals such as bison in the Fort Smith region when he received a radio message on June 30 that three probable whooping cranes (including one juvenile) had been seen in nearby Wood Buffalo Park. This vast wilderness park of more than 11 million acres had been established in the region along the Alberta-Mackenzie District boundary, largely to protect the woodland race of the bison. On his way by helicopter to visit a reported fire in an inaccessible boundary area of the park, Fuller scanned the area

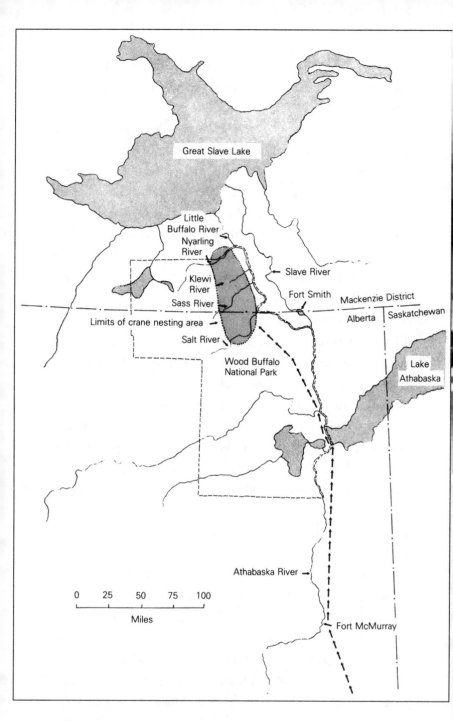

Breeding grounds (shaded) of the whooping crane in Wood Buffalo
National Park and surrounding areas. Arrows indicate the approximate
migration routes followed by the cranes during spring. Adapted from a
map in McCoy (1966).

where the cranes had been reported earlier that day, finding two adult whooping cranes. Later in the flight an additional lone adult was found some distance away near the Nyarling River. This news was quickly reported to Canadian wildlife authorities and to Robert Allen, who immediately began to arrange with Canadian and U.S. agencies for a detailed search of the area to be conducted the following year.

Thus, by 1955, the stage was finally set for the discovery of the nesting grounds, which ironically proved to be within an already protected area, namely Wood Buffalo Park. At long last, Robert Allen and Fuller's assistant, Ray Stewart, were able to reach the nesting grounds, and on May 18 Stewart and Fuller located seven whooping cranes, two of which were observed standing beside their nests. After a great deal of difficulty, on June 23 Allen reached a nesting territory by helicopter and on foot, where he was at last rewarded with a sight of the breeding pair flushing directly in front of him. After nearly a decade of effort, the greatest wildlife search in North American history was finally successfully completed. In his notes he wrote, "It has taken us 31 days and a lot of grief, but let it be known that at 2 p.m. on this 23rd day of June, we are on the ground with the Whooping Cranes! We have finally made it!" Later aerial observations that summer indicated that at least 11 adults were present in the general vicinity of northern Wood Buffalo Park in the area between the Sass and Nyarling rivers, these birds representing about six pairs and their six young. Four of the breeding pairs were found in the Sass River area, and two additional pairs plus several more apparent nonbreeders were in the Klewi River area. All told, the Wood Buffalo Park area seemingly accounted for nearly the entire North American breeding population of whooping cranes, although Allen believed that the region to the north of Great Slave Lake might at times be used by wandering nonbreeders.

During the fall of 1955 some 28 whooping cranes returned to Aransas, including 8 juveniles, the highest number recorded since the establishment of the refuge. At last the hinge of fate had apparently turned. The birds continued a painfully slow ascent in numbers, so that by 1965 there were 44 birds recorded at Aransas, and by 1975 there were 57.

During the 1960s a captive flock of whooping cranes was begun at the U.S. Fish and Wildlife's Research Center in Patuxent, Maryland. This captive population was produced by the removal of single eggs from the clutches of wild pairs nesting in Wood Buffalo Park, incubating them, and rearing the young in captivity. It was hoped that basic biological information on whooping cranes could thereby be obtained, and furthermore that offspring of these hand-reared birds might eventually be produced and released into the wild. However, this flock has never become very large, in part because of high chick mortality, and in 1988 it contained only six reproductively active females. Partly to reduce the dangers of losing part or all of this flock to disease, it is planned that a second captive flock will be developed in Canada, starting in the early 1990s. Additionally, half of the Patuxent flock of whooping cranes has recently been sent to the International Crane Foundation in Wisconsin to protect this invaluable gene pool more effectively.

In the mid-1970s the Canadian Wildlife Service and the U.S. Fish and Wildlife Service began an even more experimental program. It was designed both to increase the total whooping crane population and also to try to establish a second independent flock, whose migration route might be much shorter and less hazardous than the long route from Wood Buffalo Park to Aransas National Wildlife Refuge. Spearheaded by such people as Dr. Roderick Drewien, the idea was to remove one of the two eggs from as many whooping crane nests as possible, and to hatch some of these for captive rearing and eventual breeding at Patuxent. Other eggs would be quickly transported to Grays Lake National Wildlife Refuge in southeastern Idaho, where they would be placed in the nests of greater sandhill cranes, for hatching and foster rearing by these birds. The remaining whooping crane egg would be left in each nest for its parents to raise normally. Although this would seemingly reduce the potential productivity of the Wood Buffalo Park population, whooping cranes rarely succeed in raising two chicks, owing in part to the competition for food that occurs between the young. Thus, by having only a single chick to raise, the pair's productivity is not noticeably reduced and possibly even enhanced, owing to the greater care given by the parents to their single chick.

This innovative idea posed a great number of risks, such as the danger of the whooping cranes abandoning their nests upon disturbance, the possible damage to the eggs in transit, and the danger that the sandhill cranes would fail to accept and rear their foster offspring. There was also the risk that the young whooping cranes would not respond appropriately to their alien "parents" or, even more ominously, might "imprint" on them and grow up thinking that they too were sandhill cranes, and thereby become unable to recognize and mate with their own species.

All of these points were weighed carefully, but it was finally decided that the possible risks were outweighed by the potential benefits, and the project was initiated in 1975. The eggs were substituted by removing both of the sandhill crane eggs from an active nest and replacing them with a single whooping crane egg. At times only a single sandhill egg was removed, the second one being taken away prior to hatching. In addition to wild-taken eggs, some additional eggs were obtained from the captive flock of whooping cranes that had been gradually built up at the Patuxent Research Center. Between 1975 and 1977 a total of 45 eggs from wild whooping cranes were transported to Grays Lake, and an additional 16 eggs were sent there from the captive flock at Patuxent. The project seemed to be dogged by bad luck from the very beginning, with many losses of eggs and chicks to cold weather, nest flooding, and predation by coyotes. Of the 45 wild-taken eggs sent to Grays Lake during the first two years of the project, 35 hatched, 13 fledged, and three of the young were still alive two years after hatching. Only five of the 16 eggs from

Patuxent hatched, and none of these chicks survived to fledging. Much of the mortality of young chicks was caused by coyotes, but the young that did survive migrated as expected with their foster parents to traditional greater sandhill crane wintering areas in the Rio Grande Valley of New Mexico.

In spite of these early disappointments, additional egg transplants to Grays Lake were conducted on a yearly basis through 1988. As a result, the Grays Lake–Bosque flock of foster-raised birds slowly grew. During the same period the original Wood Buffalo–Aransas flock was also thriving, in spite of these annual egg removals (annual recruitment rates of young on arrival at Aransas averaged 10.7 percent for the period 1975–85, as compared to 11.2 percent for the prior decade). By 1988 some 289 eggs had been transplanted to Grays Lake from Wood Buffalo Park and Patuxent, and the new flock reached a maximum of 33 birds in 1985. However, since then it has rather rapidly declined, so that by the spring of 1989 it numbered only 13 individuals. Causes of this precipitous decline are uncertain, but deaths caused by powerline collisions have proven to be a serious threat to the birds, and additionally avian tuberculosis has been detected in the population. Of equal concern is the fact that none of the fostered whooping cranes have attempted to breed, or have even formed pair bonds, resurrecting fears that early exposure to alien sandhill crane parents may indeed have seriously affected the normal socialization processes and sexual tendencies of the whooper chicks. Because of the apparent failure of the Grays Lake transplant, it was decided in 1989 to abandon further egg transplants there. Instead, future efforts of this type might use a more southerly location where no migration at all need occur. The locality most likely to be chosen is Kissimmee Prairie in Osceola and Okeechobee counties of south-central Florida, where a protected resident population of Florida sandhill cranes is available to provide potential foster parents, and a relatively high level of reproductive success has been typical in recent years.

More encouragingly, the original flock of whooping cranes

wintering at Aransas reached a record total of about 150 birds by the winter of 1989–90, including some 32 breeding pairs and about 20 juveniles. By the spring of 1990 the total whooper population included 141 birds in the main Wood Buffalo–Aransas flock, 13 in the declining Grays Lake–Bosque flock, and more than 50 additional individuals in captivity, the majority of which are at the International Crane Foundation, in Wisconsin.

During the fall of 1990 a total of 155 whoopers left Wood Buffalo Park, including 14 juveniles, which were the end-product of 32 initiated nests. This generation of young birds represented an annual recruitment rate of about 9 percent, or slightly below recent averages. In order to counterbalance its annual postjuvenile mortality and continue to grow significantly, the Wood Buffalo flock's annual recruitment rate must regularly exceed 10 percent. Should favorable reproduction at Wood Buffalo Park as well as subsequent high postjuvenile survival levels of at least 90 percent continue, the hopes of Canadian and U.S. authorities of a minimum nucleus of 40 wild breeding pairs could be reached by the mid-1990s. However, the ever-present dangers of wintering-ground crowding, disease, oil spills, hurricanes, and other potential disasters to this highly vulnerable flock must always be kept in mind, and it may still be some time before whooping cranes will finally have something to whoop about.

## THE YEAR
## OF THE
## WHOOPING CRANE

*The Winter Season*

For nearly half the entire year, from about early November until mid-April, the primary flock of whooping cranes occupies traditional winter quarters along a very small part of the Texas coastline. There the birds spread out as pairs and family groups, which maintain exclusive foraging territories. These territories along salt-water fringes are established soon after fall arrival and are maintained until just prior to spring departure. At the edges of these territories the birds may threaten or even fight with one another. Within the territories the birds forage on a wide diversity of animal and plant materials, but their primary foods consist of polychaete worms, pistol shrimp, mud shrimp, blue crabs, crayfish, and razor crabs. The most important of these are the blue crabs, mud shrimp, and other crustaceans, which are abundant and easily captured. Other animal materials are probably taken as they are available, but plant foods are apparently of far less importance in the winter diet than is the case with sandhill cranes.

On the average, about 400 acres of salt flats, ponds, and estuaries comprise a single territory, which typically includes various salt-flat ponds and beach frontage along one or more of the inside bays. The male of the pair takes primary responsibility for defending the territory, and also serves as the leader in making any movements to various areas within the territory. Occasionally two pairs in adjacent territories may gradually approach one another during foraging, calling in a challenging manner as they reach the edges of their territories. Such signaling is likely to suffice, and usually the birds gradually move apart again. Exceptions to the code of territorial behavior may occur if the owners of adjoining territories are closely related, in which case an unusual degree of tolerance may be typical.

Once Robert Allen observed two males approaching each other at the very edges of their territories where these came into contact, finally facing one another while only about a yard apart. They then dropped their black primary wing feathers, raised their ornamental wing plumes (the elongated innermost secondary feathers), pointed their bills skyward, and emitted their trumpeting calls. The nearby paired females soon joined in, but the juvenile offspring of both pairs paid no apparent attention to the proceedings. After a short circling walk, with their heads held low as if foraging, the adult males again came close together.

This time each made short bowing movements and finally stood with their bills held so low that they nearly reached between their legs, and their crimson crowns were almost touching. Eventually the pairs separated and moved back into their respective territories.

This observation provides several insights into the social behavior and signals of cranes, and deserves some additional comment. First, it illustrates that much of the aggressive signaling of cranes consists of slow, often rather stately movements that emphasize the head (and especially the bare crown), the often contrastingly colored black primary feathers, and the long, ornamental inner wing feathers. Calling by one or both members of the pair may supplement these visual aspects of threat display, and probably provides a kind of emotional bonding as well as a united front for the pair when they are confronted with territorial intruders. Lowering of the head and drooping or partial spreading

of the wings are also evident in the very preliminary stages of "dancing," which thus may be seen as a more complex and apparently more elaborately "ritualized" version of what is essentially aggressive behavior.

Allen thought that the whooping crane's dance must have "an emotional basis" and that as such it probably serves to strengthen the sexual bond between the pair. However, he also believed that at times it might additionally represent a kind of generalized emotional and physical outlet, or even possibly a means of "relaxing." He also observed that such dancing (in sandhill cranes at least) seems almost involuntary in its onset, and saw no evidence that sandhill cranes engaging in such dances

during spring migration were paired. He thought that perhaps these spring dancing activities might serve as a generalized pre-nuptial display, readying the birds physiologically and psychologically for the release of actual courtship dancing when they arrive on the nesting grounds. Recent observations of color-banded whooping cranes have also indicated that dancing behavior is more frequent in subadults than in adults, suggesting that dancing may indeed play some role in pair bonding.

If crane dancing has an aggressive origin, could it possibly serve as a means of effective pair bonding? Perhaps it can, just as the "triumph ceremonies" engaged in by nearly all species of geese and swans in their pair-bonding behavior are believed to represent and be derived from redirected aggressive tendencies. Thus, by channeling potentially dangerous aggressive responses to another bird (who might represent a potential mate) into a harmless "dance," it may be possible to reduce the danger of actual fighting breaking out between them and furthermore provide a kind of mutual social stimulation. Additionally, the "unison call" that is uttered by the pair when they are excited, such as in the situation described by Allen, provides an equally important device for mutual stimulation and coordination, and indeed is probably much more important in pair bonding than is dancing behavior.

### The Spring Migration

By the second week of April, the warm southeasterly winds coming out of the Gulf of Mexico have already driven the sandhill cranes north to the Platte Valley of Nebraska, if not beyond, and many of the migratory shorebirds have also left the coast of Texas to cross the Great Plains toward their rendezvous with far-off arctic nesting grounds in Alaska and Siberia. During March the whooping cranes spend an increasing amount of time on land, especially on higher sites, where perhaps they show a gradually increasing tendency to take flight. But there is still no coalescence into spring flocks; instead the pairs and family groups remain distinct, with territorial boundaries being maintained. Then one day, perhaps with a south wind and clear skies, each family somehow makes the fateful decision to leave, and with that they take off, circle about to gain a thousand feet or more of altitude, and finally head north without hesitation.

The usual spring migration flock pattern is comprised of only two or three birds, consisting of a pair and their single surviving juvenile of the previous year. However, a surprising number of whooping cranes migrate north as single birds. A few flocks consist of from four to as many as a dozen birds, these presumably being composed of multiple family units or immature nonbreeders. When flying in groups of three or more birds, the typical flock pattern is more regularly V-shaped than is true of sandhill cranes. The older birds take the lead and the younger ones tag along behind, thereby gaining the benefits of reduced wind pressure and improved wind flow from the bird in front, relying on the lead bird to make decisions on stopping places.

Although cranes are normally rather slow-flying birds, while on migration they often remain in close formation and assume a distinctly streamlined body and wing conformation. Robert Allen judged that a typical cruising air speed for whooping cranes is 45 miles per hour, which with the aid of moderate tail winds would certainly produce ground speeds in excess of 50 miles per hour. Since the birds regularly choose following winds for migration, ground speeds are likely to be greater than observed air speeds.

The trip north from Aransas to the Platte Valley of Nebraska is done over a several-day period, especially for pairs that are leading young. Traditionally, the Platte Valley has been the most important single spring stopover area for whooping cranes, the birds remaining in the region for several days. In earlier times the birds roosted on the river at night and fed on frog and toad egg masses that they found in the numerous remaining buffalo wallows. The whoopers also seemed to favor the open areas of buffalo grass, turning over the cattle chips and feeding on the beetles that they found underneath. At least in early years, some of the birds remained in the Platte Valley until about the first of May, or after all the other spring migrants had already left except for the American white pelican (a species whose similarity to whooping cranes probably has confounded some of the earlier migration records for Nebraska).

In recent years, dewatering of the Platte River in Nebraska for irrigation and other purposes has reduced its attractiveness to whooping cranes, but it is still sufficiently important that it was designated as "critical habitat" for the species. During 1971 the U.S. Fish and Wildlife Service tried to establish a national wildlife refuge of about 15,000 acres of prime river habitat, but it alienated local landowners by trying to obtain the land through condemnation rather than using fair purchase offers. As a result, these efforts failed miserably, and it was feared that no real protection would ever be provided. However, in 1973 the National Audubon Society used an unexpected but highly serendipitous bequest of a New Jersey woman, Lillian A. Rowe, to purchase nearly 800 acres of habitat for a nature sanctuary to be established along the Platte's major channel slightly east of Gibbon, in prime crane habitat.

A few years later, in 1979, the Platte River Whooping Crane Critical Habitat Maintenance Trust was established as part of the habitat loss mitigation settlement associated with the building of Grayrocks Dam. Together with the Nature Conservancy, the Platte River Trust has identified and purchased more than 4,000 acres of crane habitat in the vicinity of Grand Island, centered on Mormon Island. The Trust's sanctuary on Mormon Island consists of a variety of wet meadows, shallow river channels with sandbars, variably vegetated islands, and riverine woodland vegetation that, together with the Rowe Sanctuary, have preserved the best of the few remaining areas of Nebraska's Platte River still usable by whooping cranes.

As important as these developments might be, they have not removed all the dangers for further dewatering. At present, plans are well along for a dam on a tributary of the upper North Platte in Wyoming (Deer Creek), which would have significant downstream effects in Nebraska that are likely to impact not only the whooping and sandhill cranes but also such regionally threatened breeding species as the least tern and piping plover. Minimum-flow requirements that are still to be established for the long-overdue relicensing of Kingsley Dam on the upper North Platte River in western Nebraska will have a critical impact on determining spring river flows during the crane migration period, and also the summer breeding season for terns and plovers. Furthermore, projected in-state agricultural projects such as the Prairie Bend and Twin Valley irrigation district projects would directly impact the water flows of the central Platte Valley. Other proposed Nebraska projects (Catherland and Landmark) would divert the Platte's water into other drainage basins for irrigation and groundwater recharge purposes.

Since the establishment of the Rowe Sanctuary and the Trust's habitat acquisition at Mormon Island, whooping cranes have been using this part of the Platte Valley with increasing frequency, usually arriving about the time that the last sandhills are departing, and remaining from a few days to a week or two. After leaving Nebraska, the birds continue on their slightly west-of-north bearing, crossing the central Dakotas, making a diagonal crossing from southeastern to northwestern Saskatchewan, and entering northeastern Alberta in the general vicinity of Fort McMurray. By now they have left the grasslands and croplands of the Great Plains, crossed the mixed hardwood-conifer forests of central Saskatchewan and Alberta, and entered the great boreal forest region that stretches from the Pacific coast of Canada to the Atlantic. They soon become swallowed up in that enormous wilderness, somehow finding their way to their traditional breeding grounds at Wood Buffalo Park. Quite possibly they encounter the Athabaska River or its tributaries shortly after entering Alberta, follow it north to Lake Athabaska, and then simply follow the Slave River to reach their nesting grounds in northern Wood Buffalo Park. Young birds remain with their parents until they arrive at the breeding grounds, after which the family bonds are at least temporarily broken. Considering that some of the cranes in the population may well be 30 years old or older, and many of

them live at least 10 years, memory of landmarks encountered during previous migrations must play an important role in the species's migratory traditions. Current data indicate that usually about three-fourths of the young whoopers banded in Canada survive their first fall migration, and for adults the annual mortality rate is around 5 to 10 percent per year.

## The Breeding Grounds

The final spring destination of the birds is a subarctic muskeg-like area of glaciated potholes, with innumerable ponds and small lakes varying in size from less than an acre to about 60 acres. All of these are quite shallow and are separated from one another by low, sandy ridges that support a dense growth of birches, willows, black spruce, and tamaracks. The ponds in turn have edges that are densely covered with bulrushes, cattails, sedges, and all of the other shoreline plants of the region. Because of the fairly recent glacial action (the area lies at the southern boundary of the Precambrian Shield, a vast area of scoured-off ancient rock that covers much of arctic Canada), the surface clay soils are of limestone-derived glacial till materials that are quite high in calcium rather than being distinctly acidic. The cranes use those ponds that are slightly on the alkaline side of neutrality, ignoring or avoiding nearby ponds that are somewhat more acidic and less rich in invertebrates. Additionally, the birds use only those potholes that are shallow enough to allow for easy foraging by wading.

Sharing the whooping crane's nesting habitat are such lake- or marsh-adapted birds as the sora, common snipe, American bittern, red-winged blackbird, and Pacific loon, shoreline-nesting sparrows such as the song sparrow and Lincoln's sparrow, and a variety of ducks including the mallard and green-winged teal. On slightly drier sites lesser yellowlegs nest, and American kestrels, dark-eyed juncos, chipping sparrows, and northern flickers all find suitable nest sites in the adjoining more wooded habitats. Wolves are not uncommon in the region, and red foxes and lynxes also occur, but of all these perhaps only the wolf might constitute a significant threat to whooping cranes or their young.

Although the area does not receive much annual precipitation, drainage is poor, and higher than normal precipitation early in the nesting season may cause nest flooding, delayed nesting, and lowered overall breeding success. On the other hand, lower than normal precipitation allows the nesting season to proceed normally, although the adults and young may have to travel farther from the nest site to forage. Regardless of the weather, there are always enough potholes of varying depths to allow for adequate foraging. The most important foods for the cranes are immature stages of dragonflies, caddisflies, mayflies, and other insects, plus some fresh-water crustaceans. Some terrestrial foods such as berries are also eaten, but much of the protein needed by the growing chicks must come from aquatic sources.

Probably because of the great food needs of adults and young, breeding territories of whooping cranes are extremely large. One estimate of 18 breeding territories found that these averaged nearly 1,900 acres each, but with considerable individual variation. Substantial seemingly unused areas often existed between adjoining nesting territories, at least in the case of the larger territories. Territories are typically used by the same pair for many years, and probably are vacated only with the death of both adults during the same year. However, the same nest site is rarely if ever used in successive years, although the birds may nest in the same marsh. Because of the low population density, there are few actual contacts with or territorial conflicts between adjoining territorial pairs, although resident birds will attack and

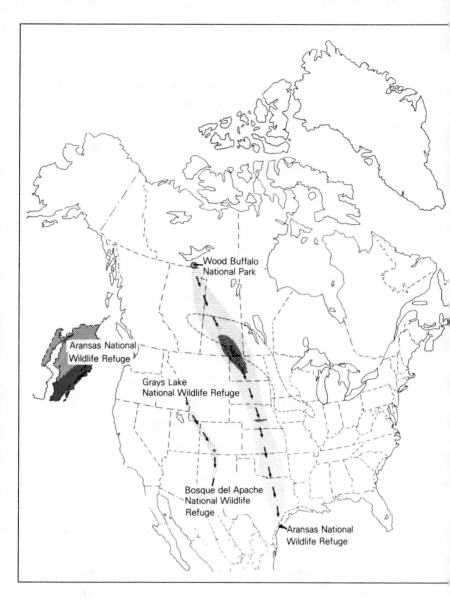

Breeding (hatched) and wintering (inked) ranges of the whooping crane, together with major migratory routes (arrows), primary migratory staging areas (cross-hatched), and total current migratory corridors (stippling). Probable historic breeding and wintering areas are shown by broken lines. The inset at left shows Aransas National Wildlife Refuge (hatched, with areas of major winter crane use cross-hatched). Modified from Johnsgard (1983).

frighten off any intruding cranes, whether these are single individuals or pairs.

As in other cranes, pair bonds in whooping cranes are essentially permanent and potentially lifelong. Although apparent pair-bonding behavior has been observed in two-year-old birds, it is likely that they usually begin nesting as five- to six-year-olds, by which time the birds probably are sufficiently experienced to establish and defend suitable nesting territories successfully. The youngest whooping crane so far known to have nested in the wild at Wood Buffalo was three years old, and the oldest was seven years old when it first nested. In spite of such delayed sexual maturity and permanent pair bonding, new mates can sometimes be taken fairly rapidly when the situation requires. Thus, when one of the adults (probably a male) of a wintering pair was lost in January to unknown causes, the surviving bird had taken on a new mate within a period of only three weeks. Likewise, during a lifetime of at least 35 years, a male called Crip was known to have had a total of five different mates, three of which were provided to him under captive conditions. After his first wild mate was shot in March 1948, he was observed with a new mate within a month. Since copulations are rarely if ever observed on the wintering grounds, it is apparent that this behavior must play little or no role in pair bond formation.

Nests are usually built along the margins of lakes or marshes, among rushes and sedges growing in water from about 8 to 18 inches in depth. The nests are huge, ranging from 2 to 5 feet in diameter, rising 8 to 19 inches above the surrounding water level, although of course changing water levels can greatly alter both the surrounding water depth and the height of the nest above water. During the period when breeding occurred in the Great Plains, nests were also found on the tops of muskrat houses and on damp prairie sites near water.

Eggs are laid at intervals of two days, and the vast majority of nests have two eggs present in completed clutches. Less than 10 percent of the nests are incubated with only a single egg present (some of these possibly resulting from egg losses), and even more rarely (in about 1–2 percent of the nests) are three eggs present. Although renesting is known to occur fairly commonly in greater sandhill cranes to the south, where there is more available time for breeding, only a few cases of apparent renesting have been found in whooping cranes following nest abandonment or other causes of nest loss.

Both sexes assist in incubation, with the male perhaps incubating somewhat more during the day and the female doing more nighttime sitting. The nonincubating bird always remains alert and fairly near the nest, ready to sound an alarm and to ward off any intruders, whether they be small or potentially serious threats to the sitting bird and the eggs. Incubation requires 33–34 days in the wild (somewhat less under conditions of artificial incubation), with the second-laid egg normally hatching two or three days after the first since incubation begins immediately after the first egg is laid.

Newly hatched whooping cranes are very similar in color to the golden-garbed chicks of the sandhills, although they are substantially larger, averaging about five ounces. They are able to swim about almost immediately after hatching, but are brooded frequently by the parents, especially during the cool nights and often unsettled weather that is typical of late June and early July in this region. During the first three weeks or so after hatching, the family remains within a mile or so of the nest. By the time they are three weeks old, the chicks weigh about four times what they did at hatching, and by six weeks they are almost 16 times their hatching weight. That weight at six weeks is again doubled by the time they are ten weeks old.

The birds move around to feed at a substantial pace following hatching, and evidently never return to their nest site. They hide in dense vegetation when necessary, although the white plumage of the adults and the rusty color of the chicks make it difficult to remain completely invisible. The young and gangly cranes, now appropriately called "colts," fledge when they are at

least 70 days old (or by no earlier than the middle of August). Until they can fly they are highly susceptible to predation by wolves and perhaps other predators. Even after they fledge they continue to be fed by their parents, especially by the female. However, by the time the young birds have fledged it is already time for them to begin the perilous and arduous fall migration southward.

When the chicks have fledged they have nearly lost their food-begging call, but have acquired a flight-intention call and an alarm call. After their voices have changed to the adult type (which typically occurs in cranes when they are nearly a year old), they will begin to utter loud "guard calls" and "location calls." The guard call is usually uttered during the collective threat behavior of adults and young toward other crane families, or toward other somewhat frightening stimuli. The location call is similar but more plaintive-sounding, and is used to help locate other cranes when the birds are visually isolated from one another. However, the important "unison call" does not appear until the second or third year of a crane's life, when it begins to assume its initial pair-bonding functions.

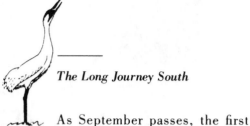

*The Long Journey South*

As September passes, the first signs of autumn are all too evident in northern Alberta. The golden fires of the aspens and birches soon flare and are quickly gone, and frost touches everything with ever sharper fingers.

The fall migration of the whooping cranes is done more as a series of waves than as a single coordinated movement. Thus, it is likely that single birds and unsuccessful pairs begin to move south early, to be followed later by family groups. Some immature or nonbreeding birds even appear in Saskatchewan by early September, while successful pairs are still tending unfledged young in Wood Buffalo Park.

The flight south is the reverse of the northward route, although stopover times along the way are likely to be rather different. Rather than the Platte, the birds are much more likely to make the grainfields and prairie marshes of western Saskatchewan their most important single fall staging area. This stopover pattern is also typical of lesser and Canadian sandhill cranes, whose routes pass through this part of Canada and whose fall stopover periods overlap with those of the whooping cranes.

By late September, family groups are starting to leave the Wood Buffalo Park area, and may join some of the whooping cranes that are still in Saskatchewan. Some whooping cranes are likely to stay in the Canadian provinces well into October, and rarely as late as November, but by mid-October they are starting to show up in North Dakota and even as far south as Nebraska and Kansas. Typically, the birds pass through Nebraska's Platte Valley between October 10 and 25 and may begin appearing on the Texas coast by late October or early November.

Fall migration thus consists of three general phases. The first of these is a rather rapid and direct flight of two or three days to Saskatchewan, a distance of more than 600 miles. The second phase consists of foraging and resting in a rather diffuse region of about 25,000 square miles of west-central Saskatchewan, where they may remain for as long as 24 days. Like the spring staging of sandhill cranes in the Platte Valley, this phase is probably of great importance to the cranes in storing needed fat reserves. The final phase consists of a fairly rapid series of flights from the staging grounds of Saskatchewan to the Aransas wintering grounds.

In recent years it has become possible to follow the migration of individual birds or family groups very accurately, inasmuch as some of the birds have been equipped with small radio transmitters that allow for tracking their every movement. For example, one family group was tracked all the way from Wood Buffalo Park to Aransas during the fall of 1981. The family left the park on October 4 and flew 175 miles to the Fort McMurray area of Alberta, where they remained for five days. On October 9 they flew another 270 miles to Reward, Saskatchewan, and there they spent 11 days. On October 20, as snow was falling, they flew 175 miles to Swift Current, Saskatchewan. On the next day they flew 150 miles to Plentywood, Montana, and 470 miles the day thereafter to Valentine, Nebraska, on the Niobrara River. On October 23, they flew about 125 additional miles to the Platte River, near Kearney. On October 24 they continued another 190 miles to Rush Center, Kansas. On October 25 they went on another 120 miles to Wanoka, Oklahoma, and on the following day flew 140 more miles to Lawton, Oklahoma. On October 27 they flew only 30 miles to the Red River, near Byers, Texas. They remained along the Red River until November 1, when they flew 230 miles to Rosebud, Texas. The next day they went an additional 178 miles to Tivoli, Texas, spending that night only some 18 miles from Aransas. The next morning they made the final short flight into Aransas, thus completing a total migration of nearly 2,300 miles in approximately a month's period. The maximum straight-line flight distance covered during any single day was 470 miles. This distance represents a rather notable feat even for the adults (representing ten hours of continuous flying at an average ground speed of 47 miles per hour), but is especially remarkable for their young offspring that had almost certainly fledged less than a month previously. During the following year a flock of five birds (a family of three plus two additional birds) made a maximum single-day flight of 510 miles in 10.8 hours, likewise averaging 47 miles per hour. A general analysis of telemetry data for the entire 2,450-mile migration route by one family resulted in a calculated average ground speed of 27 miles per hour, with an average flying time of about six hours and 28 minutes per day, excluding days of resting.

By the time the young whooping cranes have arrived at Aransas, they show a mottled mixture of white and tawny feathers, but still have only high-pitched "baby voices" and lack any bare forehead skin. Perhaps by retaining these juvenile feathers through their entire first year they are better able to maintain their strong parent-offspring bonds, and less likely to stimulate any aggressive responses from their parents or other adult cranes. Nevertheless, the young birds gradually gain a degree of independence from their parents during their first fall and winter, though they remain within their overall care and protection and follow them back north the following spring. Probably only when their parents reach their nesting territory are the young birds, now nearly a year old, finally severed from their parents' care. From then on they are increasingly forced to survive on their own.

Probably pair bonds in whooping cranes are established extremely slowly, through a delicate socialization that may begin at about two years of age; as in the sandhill crane, pair bonding is evidently a highly tentative process that may require about two years to complete. Thus actual nesting attempts may not begin for as long as two years after sexual maturity is reached, or probably usually only after females are at least five years old and often even older.

At about the same time that the whoopers are returning to Aransas, the sandhill-raised whoopers produced at Grays Lake National Wildlife Refuge of Idaho are arriving with their sandhill foster parents on their wintering areas along the Rio Grande Valley of New Mexico. While at Bosque del Apache, the whooping cranes thus keep company with large flocks of greater sandhill cranes, as well as about 50,000 or more lesser snow geese and Ross' geese, all of which share these marshlands in close proximity and relative harmony.

The increasingly large flocks of geese and cranes at Bosque have placed severe strains on the capabilities of this and other Rio Grande refuges to support them, and thus several of the New Mexico refuges have been opened to goose hunting to try to reduce and disperse the flocks. In 1988 a limited hunting season on sandhill cranes was also permitted in the Rio Grande Valley of New Mexico. So far, only a single whooping crane has been reported accidentally wounded by hunters, but other sorts of mortality factors are potentially much more serious. These threats include accidental collisions with power lines and disease transmission among the crowded and sometimes stressed birds. In the unusually cold winter of 1984–85, some 700 snow geese, 21 sandhills, and perhaps one whooping crane died of avian cholera. This highly contagious, often stress-related disease has occasionally caused tremendous losses among wild goose flocks. It could potentially eliminate the Grays Lake–Bosque whooping crane flock, which, after peaking at more than 30 birds in 1984–85, had by the spring of 1990 diminished to only 13.

Likewise, at Aransas, the wintering whooping cranes are not nearly so secure as their recent increasing numbers (155 in the winter of 1990–91) might suggest. Important coastal pond areas used by the cranes for foraging have eroded at a rate of up to about three feet per year, largely as a result of wave action caused mainly by heavy boat traffic through the adjacent Intracoastal Waterway. Similarly, the deposition of sediment spoils associated with dredging activities by the U.S. Army Corps of Engineers has affected critical crane wintering habitat, reducing it by as much as 1,150 acres in the past half century. In the process of such dredging, the Corps may well have been violating the Endangered Species Act. Legal action against the Corps by the National Audubon Society has forced it to agree to reassess the biological effects of its dredging.

Clearly, constant vigilance will be required if the species is to continue to prosper and remain a primary symbol of North American bird conservation efforts. Difficult economic and ecological choices are ahead, such as whether the Intracoastal Waterway should be relocated well outside the limits of Aransas Refuge, whether the refuge's shoreline can somehow be stabilized to resist further erosion, or whether some other less costly solution to the problem of habitat loss and environmental degradation can be found. And as the total population increases, the question of providing enough suitable wintering habitat for the birds must be addressed. Like the critical role of the endangered Platte River in the present and future well-being of sandhill cranes, whooping cranes cannot long survive without the security and winter foraging sites provided by areas such as Aransas Refuge.

Thus cranes, like all creatures including humans, are never far separated from their environments, and it does little good to increase hatching or birth rates by such impressive tricks as egg-swapping or other technological feats, if serious environmental problems provide ultimate population constraints. Like both local and worldwide human populations', their quality of life is often inversely related to the amount of crowding that must be endured. Cranes, even more than people, survive best in wilderness surroundings, and precious little of that remains today. Perhaps the question to be posed should not be whether as a society we can afford to maintain an environment that still includes enough wild and unexploited places to support a flock of wild whooping cranes, but rather whether we can afford not to.

# A GATHERING OF THE WORLD'S CRANES

**BLACK
CROWNED
CRANE**

*(Balearica pavonina)*

Black crowned cranes occur in sub-Saharan Africa from Senegal, Sierra Leone, Nigeria, and northern Cameroon on the west eastward to the upper Nile Valley. The more western populations (the "West African black crowned crane") differ slightly from those of the upper Nile valley (the "Sudan black crowned crane"); the latter race extends from about Khartoum south to Lake Turkana and east to the Ethiopian lakes. In my earlier book I treated the gray crowned crane of southern Africa as only racially distinct from the black crowned crane, but the most recent handbook on the birds of Africa regards them as a separate species, and they are so treated here. Although good data are lacking, it would appear that the West African black crowned crane is rapidly declining and perhaps numbers in the vicinity of 10,000–15,000 birds, while the Sudan black crowned crane is of uncertain status. The major remaining concentrations of the western race apparently are in the Lake Chad basin (7,000–10,000), Mauritania (2,500–3,500), and Senegambia (2,000–3,000). It has been extirpated from Sierra Leone and from much of Nigeria (where ironically it has been designated as the national bird).

Black crowned cranes differ only slightly from their gray-necked and more southerly relatives. Besides their generally darker body coloration, they also have a much smaller gular wattle, which apparently serves also as a vocal resonator. As a result, the calls of the black crowned crane are higher-pitched and less strongly resonated than those of the gray crowned crane, resulting in disyllabic notes of variable pitch. Birds of the eastern race of black crowned crane also tend to be somewhat smaller and darker than those from farther west, and they exhibit a smaller white area on the upper cheek than do most western birds. In both races the lower portion of the cheek is tinged with red.

Studies by Lawrence Walkinshaw on the cranes breeding in Nigeria indicate that the birds are highly territorial, with nesting territories ranging in size from 200 to more than 900 acres. Not only are other crowned cranes excluded from the nesting territories, but so are spur-winged geese, various ducks, and bustards. The cranes also frighten away cattle that approach the nest too closely.

Like the gray crowned crane, these birds produce clutches that average three eggs, as compared with the two-egg clutches typical of *Grus* cranes. Eggs are laid at average intervals of slightly more than one day per egg. The incubation period is 28–31 days, with the period for the last-laid eggs somewhat shorter than those for the earlier eggs in the clutch, resulting in a more clustered period of hatching. Most or all of the chicks hatch within a 24-hour period, and by the second day after hatching

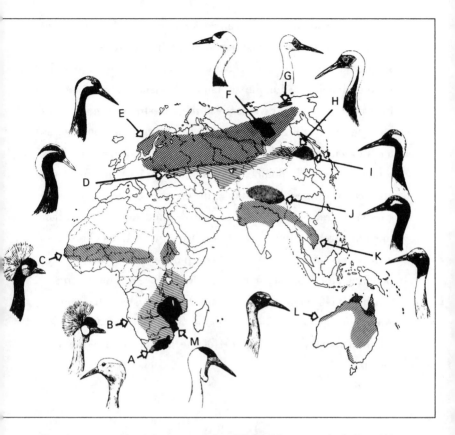

Breeding or residential ranges of the Old World cranes, including (A) blue crane, (B) gray crowned crane, (C) black crowned crane, (D) demoiselle crane, (E) Eurasian crane, (F) hooded crane, (G) Siberian crane, (H) white-naped crane, (I) Japanese crane, (J) black-necked crane, (K) sarus crane, (L) Australian crane, and (M) wattled crane.

the entire brood is able to leave the vicinity of the nest. The birds then tend to move into heavy cover, where they spend the next three or four months prior to fledging. Fledging periods are quite variable in crowned cranes. Some hand-reared West African birds did not fly until they were four months old, while some hand-reared East African birds were nearly fledged when they were only two months old. Food availability for the young must influence fledging rates in this species.

The crowned cranes, named for their tuft of golden head feathers, are the most distinctive of all living cranes, and the genus *Balearica* is usually separated from other crane genera by its placement in a unique subfamily. Evidence from molecular biology suggests that divergence of the ancestral crowned cranes and the more typical cranes may have occurred as long ago as ten million years, with the crowned crane lineage seemingly retaining a greater number of "primitive" traits than any of the other living cranes. For example, the crowned cranes are the only surviving cranes with long, prehensile hind toes, which allow them to perch well and even to roost readily in trees. The presence of this trait might suggest that primitive cranes evolved from perching-adapted ancestors. Perhaps they were similar to the brush- and swamp-dwelling limpkin *(Aramus)* of the tropical Americas, or resembled the more terrestrial and woodland-adapted trumpeters *(Psophia)* of the humid South American forests.

GRAY
CROWNED
CRANE

―――――

*(Balearica regulorum)*

This species and the related black crowned crane comprise a group of four nonoverlapping populations that are now confined to Africa south of the Sahara. They are the only living descendants of a group of cranes that once ranged widely over the world, including North America, but were eventually replaced by more advanced types of cranes. Crowned cranes have relatively short, stout bills and legs, lack elongated, decurved inner wing feathers, and also lack the highly elongated tracheal (windpipe) structure that is so characteristic of more advanced cranes. Nevertheless, they are an extremely attractive group of birds, with distinctive golden-yellow crowns, white to golden wing coverts, and grayish white to pale blue eyes.

All of the crowned cranes are associated with open country, especially favoring grasslands in the vicinity of water. Unlike the other types of cranes, they prefer to roost in elevated locations, especially large trees. However, they also at times roost in shallow water in the manner of more typical cranes. They are quite social, and outside the breeding season often occur in flocks of from a few dozen to as many as 150 birds. Like other cranes, they are strongly monogamous, and probably even during the nonbreeding season the nuclear social unit consists of the pair or family. Families remain intact for 9 or 10 months, after which the adults drive the young from their territory and prepare to nest again. At that time the young birds from the same general area tend to associate in flocks, spending much of their time foraging in fields. Crowned cranes consume a wide variety of foods, ranging from grass and grain seeds to insects, earthworms, and sometimes even crustaceans.

Like all other cranes, this species "dances," and these dances include lively leaps and bows as well as a ruffling of the long and ornamental feathers of the lower neck and breast. The wings are also often spread, exposing the beautiful contrasting upper wing coverts. The unison call of the pair is done with the birds standing in a stationary position; as the calls are uttered, the red gular sac is inflated and the head is turned slowly from side to side. The wings are not raised or moved during unison calling.

Gray crowned cranes range from South Africa north to extreme eastern Zaire, Uganda, and Kenya, with the northernmost birds separated by only a few hundred miles in northern Kenya from the eastern race of the black crowned crane. These birds are still relatively common in many areas, such as in Kenya and southern Uganda, where the population is as dense as 2–3 birds per square mile in some places. Their breeding season is highly variable, with breeding in Uganda occurring throughout the year, while in Zambia the nesting records extend from December to April, during the rainy season. Breeding during the wetter (summer) season is typical in Malawi, Zimbabwe, and South Africa, with the approximate peak of breeding there occurring in January. In South Africa they nest in open, shallow marshes, where relatively dense, tall stands of grasses and sedges occur, sufficient to hide the incubating birds effectively. Like demoiselle cranes, crowned cranes molt their major wing feathers gradually, and so never become flightless during their molting period.

## SIBERIAN CRANE

### (Bugeranus leucogeranus)

The Siberian crane had, until recently, been regarded as the second rarest crane in the world, with a total known population of a few hundred birds. In 1980 a major new wintering flock was discovered in eastern China, and new hopes were raised for the preservation of this beautiful species, which is variously called the "lily of birds" in India, the "snow wreath" in the USSR, and the "crane with black sleeves" in China. It is known to nest in only two areas of the USSR, including a small population near the confluence of the Ob and Irtysh rivers, and a second, larger eastern population nesting from the lower Kolyma River basin west to the Khroma and lower Yana rivers. There may still be other unknown breeding areas. One small flock (14 birds in 1988) winters along the southern Caspian Sea area of coastal Iran, a second flock (20 individuals in 1988) in Rajasthan, India, at the Keoladeo Ghana Sanctuary. Both of these groups represent the small western breeding population. A third flock winters in the swampy portions of northern Jiangxi Province near Poyang (Boyang) Lake, along the lower Yangtze (Changjiang) River in eastern China. This flock evidently consists of the entire breeding population of eastern Siberia, estimated in 1989–90 to consist of about 2,000 birds.

Siberian cranes are distinctly different from all the other white cranes of the world, and probably are not close relatives of any of these. Their unison call behavior and their foraging adaptations suggest affinities with the African wattled crane. Like that species, they have a rather high-pitched and somewhat gooselike voice, and the unison call ceremony is characterized by its strong wing lowering and extreme neck stretching, especially by the male. Both species also have tracheal windpipes that only slightly penetrate the front of the sternum or breastbone.

In eastern Siberia these cranes nest in arctic tundra, often in tidal flat areas or in flat and swampy depressions of ancient lakebeds that are now covered by short grasses and sedges. In western Siberia they breed in mossy marshland areas of the northern taiga forest, especially in bogs surrounded by stunted pines. During other seasons they occupy more diverse habitats, but typically occur where the birds can wade in shallow waters and where there is an abundance of aquatic plant roots. They especially like the tubers of sedges, which they can easily reach and grub out with their long and serrated bills. A small amount of animal life may be consumed at times, especially when snow cover makes plant life unavailable early in the breeding season.

The birds arrive at their Siberian nesting grounds in late May and begin to lay as soon as their nesting sites become snow-free. Their breeding territories are scattered, with individual pairs separated by distances of several miles; thus direct territorial encounters are unlikely to occur. Nests are often built at the edge of a large lake, and incubation lasts only 27–29 days, a surprisingly short period. The normal fledging period is uncertain, but a chick raised in captivity initially flew at 76 days. The wing molt occurs during this same chick-raising period in late summer, and the birds are flightless for a time. Sexual maturity is delayed in this species, judging from the limited available information on captive birds.

The Siberian crane is considered an endangered species by the ICBP and IUCN, and is also listed in the Red Book of the USSR as an endangered species. The recent discovery of the large wintering flock in China has, however, greatly improved the conservation outlook. Recent research using DNA hybridization techniques has shown this to be probably a very isolated species in a genetic sense, second only to the crowned cranes in terms of its isolation from other living species of cranes.

## WATTLED CRANE

### *(Bugeranus carunculatus)*

The distinctive wattled crane of Africa is one of the largest cranes of the world and, except for the Australian and the two crowned cranes, is the only one that is distinctly wattled. Unlike these species, however, in the wattled crane the wattle is mostly covered by feathers, with only the anterior portion bare and reddish. The front of the face (extending back to the eyes) is reddish, with wartlike papillae in both sexes, and the inner wing feathers (tertials) are greatly elongated, hiding the tail in resting birds. The wattled crane's voice is a high-pitched scream, with the male's slightly lower in pitch than the female's. In both sexes the tracheal windpipe is not deeply convoluted within the sternum.

Wattled cranes are probably the most severely threatened of the African cranes. They are now mostly limited in their distribution to the area of the upper Zambezi drainage, although in earlier times they occurred south to Cape Province and west nearly to the mouth of the Congo. Besides the Zambezi basin population, there is a small group centered in Natal and the Transvaal of South Africa, a small, outlying population in

Whooping cranes dancing.

Ovambo, northern Namibia, and a third population that once occurred locally in the highlands of Ethiopia. The current status of this last population is precarious, perhaps limited to a few pairs in Bale Mountain National Park, southeastern Ethiopia.

This crane is associated with large areas of shallow wetlands. Foraging is done by probing in wet or moist soil for the underground portions of sedges, and by consuming aquatic plants such as water lilies. It may at times also eat such prey as frogs and snakes. Thus, the wattled crane forages in much the same manner as the Siberian crane, a species that is considered by some to be the wattled crane's nearest relative.

The breeding season of the wattled crane extends throughout the year in Natal and Zambia, while in Ethiopia and Malawi the

Whooping cranes landing.

breeding records range from May to October. Breeding territories include nest sites that are situated in open grassy or sedge-covered marshes up to a meter in depth, with high, dense emergent vegetation. The nest is a substantial pile of such vegetation, with the area around the nest site well stripped of growing plants. Wattled cranes have the smallest average clutch size of any cranes, averaging only 1.5 eggs per clutch. Additionally they have a very long incubation period of 33 to 40 days, and an extremely long fledging period of 103 to 148 days. This long fledging period would seem to put the young at considerable risk from predation. Perhaps for such reasons, the reproductive success of wattled cranes appears to be very low, with fledged young usually comprising less than 5 percent of the postbreeding populations.

The total population of wattled cranes is probably 4,000 to 6,000 birds, placing it near the top of the list of crane species that deserve special conservation attention. Only recently has it been added to the list of the world's endangered crane species. Zambia and Botswana support most of this species's remaining populations, but it may be severely endangered in South Africa as well as in Ethiopia. Several proposed or current development projects in the heart of the wattled crane's primary remaining range in the upper Zambezi drainage make its conservation needs a particulary urgent matter.

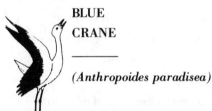

## BLUE CRANE

*(Anthropoides paradisea)*

This beautiful species, sometimes called the Stanley's crane or paradise crane, is one of only two cranes to have been designated as a country's "national bird" (by the Republic of South Africa). It also has one of the most restricted distributions of any crane, being essentially limited to South Africa including Swaziland and Lesotho, plus a small and isolated population near the Etosha Pan of Namibia. The blue crane is a close relative of the demoiselle crane, and, like it, is particularly adapted to arid grasslands. It is especially characteristic of grass-covered hills and valleys with only scattered trees, where grassy cover is thick and short. In Natal, the birds breed in highland "bergveld" areas between 3,300 and 6,500 feet elevation. There the climate is temperate, and most of the precipitation occurs during the summer months, often in the form of hailstorms. During the cold and dry winter season, the birds move to lower elevations.

Blue cranes have short and moderately pointed bills and do most of their foraging from the ground surface or from low vegetation. They have not been found to dig for foods with their bills, nor to forage in water, although nighttime roosting in water sometimes occurs among wintering flocks. At that time of year, they are quite gregarious and may form flocks of up to 300 birds. At that season, they may also forage among herds of ungulates such as springbok antelopes, with which they form an integrated society, the ever-alert cranes sometimes warning the antelopes of possible danger.

Blue cranes are territorial, and although they are too small to expel wattled cranes from their nesting territories, they do not hesitate to attack cattle or most species of birds that too closely approach their nests. When humans approach, they usually simply retreat, though they may call, dance, or circle the intruder with their wings outstretched.

The breeding season of blue cranes is limited to the summer period between October and March, with a peak of egg records in December. Nests are placed near water, although in short-grass foothill habitats they may be placed in quite dry locations. There are consistently two eggs in the clutch, which are laid from one to three days apart. The incubation period lasts from 30 to 33 days, and the hatching of the eggs is relatively synchronous, although at times the chicks are hatched on successive days.

Shortly after the last chick has hatched, the nest is abandoned, and the family gradually moves away from the nest site. There is a considerable variation in the rate of chick growth and resulting fledging, with some wild birds fledging in less than four months and others not until they are about six months old. The young remain with their parents until the following breeding season, when they are chased away from the breeding territory by their parents.

In spite of its small total range, this species is still fairly common locally, no doubt in part because of its special protected status as the national bird of South Africa. No accurate estimates of its population are available, but it is believed to number between 4,000 and 10,000 birds, as compared to about 15,000 birds only 15 years ago. Poisoning associated with farming activities is believed to be the major cause of this precipitous population decline in recent years.

## DEMOISELLE CRANE

*(Anthropoides virgo)*

This smallest, "damsel-like" and most elegant of all the cranes of the world is a relative of the blue crane. Like that species it has a fully feathered crown, distinctly elongated inner wing feathers, and a somewhat shaggy breast. It is adapted to a dry upland and grass-dominated environment. Indeed, the demoiselle crane is perhaps the most arid-adapted of all the cranes. It has a wide distribution that once included northwestern Africa (Algeria, Tunisia, and eastern Morocco) but is now restricted to the southern Ukraine and Crimea through southeastern Russia, the steppes of Central Asia, Mongolia, and northeastern China. Its breeding densities across this broad range are low, but it winters in large numbers in northern Africa (Lake Chad east to the Nile Valley) and in the Indian subcontinent. In Kazakhstan and Central Asia, the demoiselle has declined over much of its original and now rapidly disappearing steppe range, but to a limited degree has begun to nest in agriculturally modified areas.

Throughout its breeding range the demoiselle crane occurs in steppelike to semidesert habitats. It moves into marshes and swamps only for foraging or roosting. The birds nevertheless prefer to nest no more than about a mile from water, and nests often are located within a few hundred yards of it. During the winter period flocks gather in rice paddies, along the margins of shallow monsoon-dependent wetlands ("jheels") and reservoirs ("tanks"), and in other open and variably moist habitats. Roosts are often located along the sandbars of large rivers or the margins of shallow ponds, as in sandhill cranes.

The demoiselle is a gregarious crane, at least on its wintering grounds, and flocks numbering in the thousands of birds have been reported during that season. The birds commonly mix with Eurasian cranes on wintering areas, and may forage or roost with them in large, mixed flocks. Dancing has been observed among wintering birds as well as among migrating spring and fall flocks. Dancing by these small cranes is highly animated, quick and graceful, with balletlike movements, "onlooker" birds sometimes forming a loose ring around the displaying individuals.

This is a spring-nesting species, with eggs in the USSR being laid primarily during April and May, but sometimes as late as June in Siberia. Almost invariably two eggs are laid, and these are spotted and colored in such a way as to blend very well with the background. The nest is also located where small stones are present. No plant materials are added, so the surrounding stones help provide visual camouflage. The incubation period is 27 to 29 days, the shortest incubation period of all cranes. Most of the incubation is done by the female. The fledging period is 55 to 65 days, which is also extremely short for cranes. During the fledging period, the adults do not become flightless, but instead lose their flight feathers gradually, and continue to replace them during the fall migration. This gradual molt pattern may reflect an adaptation to dryland breeding, where bodies of water in which flightless birds might escape terrestrial predators are rare or lacking.

It is not known with certainty how long family bonds last, but probably they persist through the first winter of life, with the young leaving their parents when slightly less than a year of age. This is an unusually brief period of immaturity for cranes, most species of which are unlikely to breed until they are at least three years old.

## AUSTRALIAN CRANE

*(Grus rubicundus)*

This crane, which in Australia is often called the brolga (a corruption of an aboriginal name) or the native companion, is a close relative to the sarus crane. Both are tall, long-billed, predominantly grayish birds, with a head that is mostly bare in adults. However, the Australian crane is feathered somewhat farther up the neck, and it has a more distinct wattle or dewlap on the throat, as well as blackish rather than reddish legs. Both species utter strong, resonating calls; during the unison call, displaying males of both species strongly arch their wings and throw back their head and neck to a fully vertical position. In the Australian crane the unison calls are somewhat stronger and lower in pitch than are those of the sarus. Mixed pairings of Australian and sarus cranes do sometimes occur, even under wild conditions, and natural hybridization has been reported in the northern portions of Australia. In this area the Australian crane has long been abundant, but the sarus crane has only recently appeared and begun to colonize the region.

The Australian crane is widespread over the northern portions of Australia and occurs locally as far south as southern Victoria. The largest numbers and densest concentrations are found in Queensland, especially in the region between the Waverly Plains and Rocky River. There the birds seek out freshwater swamps that are dominated by *Eleocharis* sedges, on the tubers of which the cranes forage. For most of the year these tubers, locally called "bulkuru," comprise the species's primary food, but in some areas other sedges are also consumed. Various grain crops and some insects or other invertebrates may also be eaten.

Nesting in this species is timed to coincide with the wet season, which in northern Australia usually begins in December. With the onset of the rainy period, there is lowland flooding and filling of seasonal swamps and lagoons. When this occurs, nesting begins immediately, and normally the chicks have already hatched by the time the lagoons begin to dry up once again. At this time there is a gradual movement of adults and young back to the permanent coastal marshes, where some nesting also occurs. The length and severity of the dry season varies considerably from year to year, and so there are considerable variations in the seasonal movements of the cranes.

Incubation in this species lasts from 28 to 36 days, and nearly all of the nests of wild birds contain two eggs. The young are fledged by about 14 weeks of age and remain with their parents until they are almost three years old, although of course they are evicted from their parents' territory during the nesting season. The length of the breeding season in these tropical areas may sometimes permit as many as two renesting efforts in the event of early nesting failures.

## SARUS CRANE

*(Grus antigone)*

The sarus crane is the tallest of the world's cranes and is also one of the heaviest, with adult males standing nearly six feet in height and averaging more than 18 pounds. The birds range widely over the Indian peninsula, and at least originally also ranged over much of Indochina and even reached the Philippines. In the last few decades, they have managed to reach and colonize a rather large area in northern Australia (Northern Territory and northern Queensland), but they have apparently been extirpated from Luzon and a substantial portion of Indochina. They remain common in northern India, where the Hindus consider them sacred, and where perhaps they served as the original basis for the mythical garuda bird.

As in the Australian crane, most of the head and upper neck are bare of feathers in the adult sarus crane, and, except for the grayish crown, the entire head region is a startling flesh red. The vernacular name sarus is of Hindi origin. Linnaeus gave the sarus crane the specific name *antigone*, in reference to the daughter of Oedipus.

In northern India these birds are associated with a wide variety of wetland habitats, most of which are seasonal wetlands, flooded during the monsoon period. The arrival of the monsoon rains sets off breeding, but during years when there is no lowland flooding, there may be no nesting. During nonbreeding periods the birds flock to a limited degree, although flock sizes of more than 100 birds are rare. The cranes are omnivorous, consuming not only a wide array of plant materials but also animal foods that range in size from grasshoppers to moderately large water snakes.

Territorial activity begins within as little as a week after the start of the rainy season. At that time, flocks disperse and pairs begin to defend areas that range in size from 100 to 150 acres. The nests are constructed in shallow water, consisting of large heaps of vegetation that are placed among stumps or other supporting structures. Two eggs are laid, and they are incubated from 31 to 35 days. The majority of the incubation is done by the female, while the male assumes the responsibility of watching for possible danger. The adults are large enough to prevent nearly all possible enemies from approaching the nest site, including the numerous raptors that are common in the region.

The young chicks are led away from the nest after a few days. They require about 90 days to attain fledging. The young remain with their parents for about ten months, at which time the adults usually begin breeding again. Although the adults are known to undergo a flightless molt during the time that the young are being reared, little is known of its duration. The population of the western race of sarus crane is apparently doing fairly well, but the eastern race may be seriously threatened on the Asian mainland. At Tram Chin Reserve, Vietnam, the 1989–90 winter count revealed only 800 birds. In northern Australia, however, the sarus crane is prospering, and, because of its ecological advantages over the Australian crane, is rapidly increasing and may well become the dominant crane species there in a few years.

## WHITE-NAPED CRANE

### (Grus vipio)

The white-naped crane is well named; it is the only white-headed crane that has a red facial patch extending far enough backward to encompass the ear opening, and the only one that has a dark grayish stripe extending up the side of the neck to terminate at a point slightly behind the bare facial region. This species has been classified as "vulnerable" by the ICBP, and it is primarily limited in its breeding range to the USSR, where it is regarded as being very rare, with decreasing numbers. It is known to breed in small numbers at several different Soviet locations, including the middle Amur River basin and the Ussuri River valley, and at least until recently it nested commonly along the shore of Lake Khanka, in the upper Ussuri basin. It is not adequately protected in some of these areas. Perhaps Mongolia actually supports the largest area of potential breeding habitat. The total world population in 1990 numbered some 4,500–5,000 birds.

During the winter, this crane is found in eastern China, Korea (mainly near the Demilitarized Zone), and in Japan, where it occupies a restricted area in Izumi and Akune districts of southwestern Kyushu. It is in the last-named area that the best opportunities for censusing exist, and in 1990 there were about

1,500 birds present on Kyushu. The concentration of this many birds, as well as substantial numbers of hooded cranes, has caused many local problems of crop damage by the cranes, and sightseers have had disruptive effects on both the local residents and the cranes. It is vital that Japan manage its rare wintering cranes in such a way as to take into account all of these sometimes conflicting interests.

Currently the Demilitarized Zone of Korea provides a fortuitous refuge for migrating and sometimes overwintering cranes, but this is a situation that might change without advance warning. The establishment by the Chinese Ministry of Forestry of a nature preserve at Poyang Lake, Jiangxi Province, has not only been of critical importance to the Siberian crane, but has also been of great value to the white-naped crane; in 1989–90 nearly 3,000 white-naped cranes were counted there during the winter period. Additionally the USSR has included this crane in its Red Book of threatened and endangered species, and has been making strong efforts to protect it and its breeding habitats.

Incubating whooping cranes.

The birds nest in grassy or swampy areas of wide river valleys, or in lake depressions in steppe or forest-steppe habitats. Their wintering habitats are primarily brackish marshlands and rice paddies, with nearby roosting sites on salt marshes, mud flats, or the edges and sandbars of shallow lakes. Preservation of adequate areas of both of these habitat types, which are often influenced by agricultural interests, will be necessary for the continued survival of this beautiful crane.

## JAPANESE CRANE

### (Grus japonensis)

This marvelous Asian crane is also known by a variety of English names, including Manchurian crane and red-crowned crane, but none of these is particularly suitable for such a magnificent bird. It is quite possibly the most beautiful of all cranes, with a snow-white plumage that is set off by a red crown and jet-black flight feathers, plus similar black feathers on its head and upper neck, but with a contrasting white nape and hindneck. Adults are the heaviest of all crane species, with males weighing as much as 25 pounds during autumn. Like the similar whooping crane, the Japanese crane has an extremely loud voice that may easily carry a mile or more under favorable conditions.

The Japanese crane is considered "vulnerable" in the ICBP's Red Data Book, and its population in the 1970s was believed to number less than 500 birds. The majority of these were then restricted to Hokkaido, Japan, where a relict resident population survives. The remaining birds breed in the Amur and Ussuri basins of the USSR and adjoining northeastern China, and winter along coastal China and in the vicinity of the Demilitarized Zone of Korea.

The Hokkaido population of the Japanese crane was for a time believed to be extinct, but a small group of birds was found to be nesting near Kushiro in 1924. Although given protection, this population increased only slowly until the 1950s, when supplemental feeding during winter was begun. By the mid-1960s the population had reached 200 birds, and it has hovered around 200–400 total birds ever since. On the Asian mainland the species's primary breeding area is in the vicinity of Lake Khanka, where it was first discovered nesting more than a century ago. However, its habitat there has become greatly restricted in recent years, and perhaps now it is more common farther west, in the drainage of the Sungari and Nun rivers of China. A good part of this breeding flock winters in Korea, mainly around the Demilitarized Zone. The other important wintering area for the mainland population is along the coast of China's Jiangsu Province, near the mouth of the Yangtze River. Yancheng Nature Reserve in that area now perhaps provides the most critical winter habitat for the Japanese crane in China.

Apparently the preferred foods of this species are quite similar to those of the whooping crane, and include a substantial amount of animal materials. Much of its food is thus obtained by wading and probing for aquatic animals such as snails, crabs, fish, and the like. However, it also relies heavily on artificial feeding (of corn) during winter in Hokkaido.

Although the Japanese must be greatly admired for their efforts at saving this crane on Hokkaido, and in protecting a part of Kushiro Marsh from destruction, it is apparent now that the population there is essentially saturated. If the species's world population is to grow more there must be other areas of suitable breeding habitats preserved for it, especially on the mainland. As of 1990 its total known winter population was probably about 1,300 birds, including about 450 in Japan, nearly 700 in China, and about 150 in Korea, making it perhaps the second-rarest crane in the world.

## HOODED CRANE

*(Grus monachus)*

This rather small species of crane is sometimes known by its Latin-based name of "monk crane," in reference to the white "hood" that contrasts with an otherwise dark gray to blackish body. It is a species that has been classified as "threatened" by the ICBP and the best estimates are that something close to 8,000 birds were alive in 1990, based on counts of wintering birds in Honshu and Kyushu, Japan, and in China.

In spite of this moderately comforting population size, the breeding grounds of the hooded crane are almost completely undocumented, and only a very few actual breeding records exist for it. For a long time it was believed that the species bred in the vicinity of Lake Baikal and the regions to the west, but this was partly based on the erroneous identification of a nest found in the early 1900s, and on wrongly identified eggs that had been found in the vicinity of Tomsk. In fact, it was not until 1974 that the first documented nesting of the hooded crane was obtained, in the Bikin River area of the Ussuri basin, well to the east of the area previously suspected. Similar studies in the Vilyuy basin of Siberia during the early 1970s indicated that regular nesting occurs in that large but little-studied region. In both areas the birds have been found to prefer mossy hammocks or damp moors in boggy larch forests, at altitudes of about 600 to 2,500 feet. Other areas where the birds have been seen during summer, such as the open steppes and forest-steppes of Transbaikalia, are apparently used only by nonbreeding birds. So far as is known, nearly all breeding is limited to the USSR. However, a small amount of nesting habitat may occur in northeastern China, very near the Siberian border. Nesting on Sakhalin Island is also possible but still unproven.

Even in the known breeding areas, nesting densities are apparently extremely low. Evidently these densities are influenced by the availability of mossy bogs of the proper size and with suitable degrees of visual and acoustic isolation from other cranes as well as from human disturbance. Although little is known of breeding-season foods, studies on wintering birds suggest a high percentage of vegetable materials in their diet, and thus it seems likely that hooded cranes are much like Eurasian cranes in their general dietary needs. Indeed, a few instances of wild hybridization between hooded and Eurasian cranes have been reported, suggesting that these are fairly closely related species that perhaps maintain their reproductive isolation primarily by habitat preference differences during the breeding season.

This is one of the species of cranes that can be preserved from extinction only by the cooperation of several nations, partic-

Greater sandhill crane chick at two weeks.

ularly the USSR, China, Korea, and Japan. Already some important wintering areas in Korea and eastern China have apparently been abandoned, and the wintering grounds in Japan are extremely localized. In 1989–90 about 7,200 birds were wintering at Izumi, Japan, and another 400 were counted at various wetland localities in China, including Shengjin (Anhui Province), Poyang (Jiangxi Province), Longgan (Hubei Province), and Dongting (Hunan Province).

## BLACK-NECKED CRANE

*(Grus nigricollis)*

This is the least-studied of all the cranes in the world, and the species that has most rarely been maintained in captivity or even observed in the wild. Indeed, it had been represented in European bird collections only once, when Jean Delacour brought some birds to France in the mid-1920s, but breeding was never achieved. From then until the 1980s the species has been observed only by visitors to the Himalayas, or by those who have visited some of the Chinese zoos, where it has been on display since the 1960s. It was not until 1985 that the species was first brought alive into the United States, and soon thereafter to Germany.

The breeding grounds of the black-necked crane are high in the Himalayas, at elevations of about 13,000 to 15,000 feet, where tundralike marshes occur around lake edges and on lake islands. In such locations there are grass- or sedge-dominated areas that have a relatively abundant aquatic life, and grassy mounds in shallow lakes or ponds exist and serve as nesting sites. There seem to be few mammalian predators present in these bleak habitats and high elevations, and likewise there are few raptors of significance. Furthermore, the island nests are constructed well out of reach of humans and most terrestrial predators.

The breeding season is fairly short at these high elevations, probably confined to the period between late May and August. Evidently there is a fall migration out of the region in October, with many of the cranes wintering to the southeast, in the Yunnan-Guizhou regions of China or in southern Tibet, with a limited amount of wintering also occurring in Bhutan and at least formerly in Assam and Vietnam. During winter the birds mix to some degree with Eurasian cranes, and apparently have rather similar ecological requirements to that species's. However, limited information suggests that they have a greater preference for foraging in marshes and other wetlands, and thus may feed to a larger degree on animal materials.

Nearly all of the breeding range of this species is confined to Tibet, and at least until the Chinese influence became strong in the 1950s they were effectively protected by the Tibetans' sacred treatment of all animals. The current situation is not clear, but the Chinese government is currently affording the black-necked crane its highest level of official protection, and has established one sanctuary specifically for it. It is also the subject of biological studies by the Chinese, and Bhutan has recently established two protected areas on its wintering grounds.

It is hoped that eventually a breeding group can be established in North America or Europe. The International Crane Foundation received a pair of black-necked cranes from China in 1985, but unfortunately one of the birds died shortly after arrival. It was replaced in 1988, and a chick was successfully hatched and raised in 1990. One pair in the Beijing Zoo had previously bred successfully in 1987, the first time this species has ever been hatched in captivity, and another pair bred at Vogelpark Walsrode, Germany, in 1990.

Although considered to be of "indeterminate" population status by the ICBP, with population estimates ranging from 500 to as many as 10,000 individuals, it is believed by the International Crane Foundation to be perhaps the third-rarest species of crane in the world. The ICF regards the species as seriously endangered, with only about 1,600 birds actually known to exist in 1990. Agricultural development of winter habitats, poisoning by farmers in the People's Republic of China, and hunting in Tibet are believed to be the most serious factors affecting its survival.

## EURASIAN CRANE

*(Grus grus)*

This is the "common crane" of Europe, and the one having the broadest breeding distribution of any of the Old World cranes. Currently it breeds from Scandinavia on the west to at least the Kolyma River and probably the Okhotsk Sea on the east, and locally south to Germany, Poland, possibly to Romania, and to Turkey, the steppes of Russian and Chinese Turkestan, and northeastern China. Wintering areas are similarly vast and include Spain, northern Africa, Turkey, the Persian Gulf region, northern India, and from southern China south to Indochina.

Like the fairly closely related sandhill crane, this species is mostly gray, but unlike the sandhill crane it has a white stripe extending from the cheeks back toward the hindneck, and a mostly black face, foreneck, and nape. It is only moderately large, with adults weighing about 10 to 13 pounds, or about the size of a greater sandhill crane. Its voice is similarly loud and resonant, but not so penetrating as those of the whooping or Japanese cranes. During the unison call both sexes raise their curved tertial feathers, lower their primaries, and stretch their necks to the vertical.

Although the Eurasian crane has slowly lost breeding areas in the westernmost parts of its range, it is still moderately common in Scandinavia, and especially so in the USSR, which may support 60,000 to 100,000 birds. No efforts have been made to survey the entire world population of Eurasian cranes, but some 15,000 to 20,000 winter on the Iberian Peninsula, which along with the wintering cranes of northwestern Africa probably include nearly all of the European and Scandinavian breeding populations. In some years cranes winter south to Tunisia (about 10,000 estimated in 1979), and others that presumably are of Russian origin winter regularly in the river valleys of Sudan and the Ethiopian highlands. Much of the Siberian population apparently funnels into the Indian subcontinent, while those from the easternmost regions of Siberia migrate south to southeastern China. Unfortunately almost nothing is known of the population sizes of these components.

Judging from studies in Scandinavia, the breeding territories of Eurasian cranes are very large, and may range from about 125 to more than 1,000 acres. Nests of adjacent pairs are often as far as 10 miles apart and are rarely as close as a mile apart. As with other cranes, visual isolation is apparently a major factor in influencing minimal territory size, and thus an interspersion of wooded areas between suitable nesting marshes tends to facilitate greater nesting densities.

# EPILOGUE

The world's human populations have doubled from about three billion to roughly six billion in the past 35 years or so. Meanwhile, the number of crane species that have been classified as "endangered" has more than trebled since 1966, the year the first edition of the Red Data Book of rare and endangered bird species was published. It has increased from two species (whooping and Siberian) and one additional race (the Mississippi race of the sandhill crane) to nine species plus two sandhill crane races. The tabular summary of the generally rather bleak situation facing the world's cranes is thus presented here as a mere "snapshot in time," in full awareness that it will soon be outdated and of historical interest only. We humans are now in a critical moment in time, when fate is forcing us to make decisions affecting not only our own survival but also that of all the other creatures that happen to be currently sharing our planet with us, whether they be great or small, grand or obscure, "useful" or "useless."

Aldo Leopold was acutely aware of the role of time in the events of both humans and nature; his parable of the "good oak" in *A Sand County Almanac* provides a keen sense of Wisconsin ecological history. He knew too the terrible silence of a now-deserted place that once gave shelter and sustenance to wild creatures. As he said so eloquently, "The sadness discernible in some marshes arises, perhaps, from their once having harbored cranes. Now they stand humbled, adrift in history."

We humans, as a species, are also captives of history. Caught up in our daily worries and repetitive patterns of existence, the sounds of migrating cranes or geese overhead are often not heard at all above the noises of the city, or, if heard, are recognized by only a few. Similarly, we all too often cannot fully sense the slowly rising tide of our diverse ecological crises, until we begin to realize that we are in real danger of losing control of our own destiny. From the subtle but pervasive influence of the ozone layer on global climate patterns, or the far-removed effects of acid rain, we are perhaps like that unfortunate apocryphal frog that, unaware of the slowly heating water surrounding it, failed to make its escape before it was fatally boiled alive. By the time that most politicians, as well as the general public, are convinced that a real ecological crisis exists, it may well also be too late to do anything to avert it.

Since the last world war, we as a North American society have generally worried primarily about the Communists, the Bomb, and the prospect of dying suddenly in a nuclear holocaust. With the ending of the Cold War, perhaps we can now turn our attention to the area that is far more of a threat to our continued existence, that of population control, averting the wholesale ravaging of natural habitats and salvaging at least a minimal sample of the world's habitats and biodiversity. Let us also hope that a small part of that salvaged biodiversity might be represented by cranes. If I had to choose between never hearing an angelic chorus or never again hearing wild cranes, I would most certainly choose the cranes. I have indeed often wondered if the angels that were "heard on high" above Bethlehem were not really migrating Eurasian cranes—at least that's a pleasant thought to contemplate.

## A SUMMARY OF THE CRANES OF THE WORLD AND THEIR STATUS

| Species and Subspecies | Approximate 1990 World Population | Major Breeding Areas | Population Status[1] |
|---|---|---|---|
| Black Crowned Crane | | | |
|   West African race | 10,000–15,000 | Senegambia, Lake Chad area | Declining |
|   Sudan race | 50,000–70,000 | Upper Nile basin | Stable |
| Gray Crowned Crane | 100,000 + | Southern and eastern Africa | Unstable |
| Siberian Crane | ca. 2,000 | USSR | Endangered |
| Wattled Crane | 4,000–6,000 | Botswana, Zambia | Endangered |
| Blue Crane | 6,000–10,000 | South Africa | Declining |
| Demoiselle Crane | 50,000–60,000 | USSR | Declining |
| Australian Crane | 15,000–20,000 | Australia | Stable? |
| Sarus Crane | | | |
|   Western Sarus | ca. 25,000 | India | Declining |
|   Eastern Sarus | few thousand | Asia, Australia | Threatened |
| White-naped Crane | 4,500–5,000 | USSR | Endangered |
| Whooping Crane | 150 (in wild) | Canada | Endangered |
| Sandhill Crane | | | |
|   Lesser Sandhill | ca. 500,000 | Alaska, USSR | Stable |
|   Canadian Sandhill | few thousand | Canada | Stable? |
|   Greater Sandhill | 35,000–40,000 | USA | Increasing |
|   Florida Sandhill | ca. 4,000 | Florida | Stable |
|   Mississippi Sandhill | ca. 50 | Mississippi | Endangered |
|   Cuban Sandhill | ca. 50 | Isle of Pines | Endangered |
| Japanese Crane | 1,000–1,300 | China, Japan | Endangered |
| Hooded Crane | ca. 8,000 | USSR | Endangered |
| Black-necked Crane | ca. 1,600 | Tibet | Endangered |
| Eurasian Crane | 100,000–150,000 | USSR, Scandinavia | Stable? |

[1]Status based on U.S. Fish and Wildlife Service classification (1986), and population trends as judged by George Archibald (personal communication).

# REFERENCES

Allen, Robert P. 1952. The Whooping Crane. New York: National Audubon Society, Research Report No. 3.

Allen, Robert P. 1956. A report on the whooping crane's northern breeding grounds. A supplement to the Research Report No. 3. New York: National Audubon Society.

Archibald, George W. 1975. The unison call of cranes as a useful taxonomic tool. Ph.D. dissertation, Cornell University, Ithaca.

Archibald, George W., and R. F. Pasquier (eds.). 1987. Proceedings of the 1983 International Crane Workshop, Bharatpur, India. Baraboo, Wisconsin: International Crane Foundation. (Available from ICF, Baraboo, Wisc. 53913, $18.00.)

Bent, Arthur C. 1926. Life histories of North American marsh birds. U.S. National Museum Bulletin 135:1–490.

Bishop, Mary A. 1984. The dynamics of subadult flocks of whooping cranes wintering in Texas, 1978–79 through 1982–83. M.S. thesis, Texas A & M University, College Station.

Boise, Cheryl M. 1977. Breeding biology of the lesser sandhill crane (*Grus canadensis canadensis* L.) on the Yukon-Kuskokwim Delta, Alaska. M.S. thesis, University of Alaska, College.

Britton, Dorothy, and Tsuneo Hayashida. 1981. The Japanese Crane: Bird of Happiness. Tokyo: Kodansha International.

Conant, Bruce, James King, and Harold Hansen. 1985. Sandhill cranes in Alaska: a population survey: 1957–1985. *American Birds* 39:855–858.

Currier, Paul J., G. R. Lingle, and J. G. VanDerwalker. 1985. Migratory bird habitat on the Platte and North Platte Rivers in Nebraska. Grand Island: Platte River Whooping Crane Critical Habitat Maintenance Trust.

Doughty, Robin W. 1989. Return of the Whooping Crane. Austin: University of Texas Press.

Drewien, Roderick C. 1973. Ecology of Rocky Mountain greater sandhill cranes. Ph.D. dissertation, University of Idaho, Moscow.

Drewien, Roderick C., and Ernie Kuyt. 1979. Teamwork helps the whooping crane. *National Geographic* 155(5):680–692.

Drewien, Roderick C., and James C. Lewis. 1987. Status and distribution of cranes of North America. In Archibald and Pasquier, 1987, pp. 469–477.

Farrar, Jon. 1985. Partners on the Platte. *Nature Conservancy News* 35:13–18.

Farrar, Jon. 1989. Lillian Annette Rowe Sanctuary: way-station on the Platte. *Nebraskaland* 67(2):18–34.

Farrar, Jon, and K. Bouc. Undated. Sandhill cranes: wings over the Platte. Lincoln: Nebraska Game & Parks Commission. 16 pp.

Gruchow, Paul. 1989. The ancient faith of cranes. *Audubon Magazine* 91 (3):40–54. (Part of special issue of *Audubon Magazine* primarily devoted to the Platte River.)

Harris, James H. (ed.). In press. Proceedings of the 1987 International Crane Workshop. Baraboo, Wisconsin: International Crane Foundation. (Cranes of China.)

Harrison, George H. 1978. Crane saviors of Baraboo. *Audubon* 80(3):25–30. (Work of International Crane Foundation.)

Howe, Marshall A. 1989. Migration of radio-tagged whooping cranes from the Aransas-Wood Buffalo population: patterns of habitat use, behavior and survival. Washington, D.C.: U.S. Fish and Wildlife Service Report 20:1–20.

Johnsgard, Paul A. 1973. How many cranes make a skyfull? *Animals* (December), 532–539. (Sandhill crane populations.)

Johnsgard, Paul A. 1981. Those of the Gray Wind: The Sandhill Cranes. New York: St. Martin's Press. Reprinted by University of Nebraska Press, Lincoln, 1986.

Johnsgard, Paul A. 1982. Teton Wildlife: Observations by a Naturalist. Boulder: Colorado University Press. (Greater sandhill cranes.)

Johnsgard, Paul A. 1982. Whooper recount. *Natural History* 91(2):70–75. (Population trends in whooping cranes.)

Johnsgard, Paul A. 1983. Cranes of the World. Lincoln: University of Nebraska Press.

Johnsgard, Paul A. 1983. The Platte: a river of birds. *Nature Conservancy News* 33:6–10.

Johnsgard, Paul A. 1984. The Platte: Channels in Time. Lincoln: University of Nebraska Press.

Johnson, Aubrey S. 1987. Will Bosque's whoopers make it? *Defenders* 62(1):20–27. (Threats to whoopers from hunting, collisions with power lines, and avian cholera.)

Karjewski, C. 1988. Phylogenetic relationships among cranes (Aves: Gruiformes) based on DNA hybridization. *American Zoologist* 28:172A. (Abstract.)

Kessel, Brina. 1984. Migration of sandhill cranes, *Grus canadensis*, in east-central Alaska, with routes through Alaska and western Canada. *Canadian Field-Naturalist* 98:279–292.

King, Warren B. (ed.). 1981. Endangered Birds of the World. The ICBP Bird

Red Data Book. Washington, D.C.: Smithsonian Institution Press and International Council for Bird Preservation. (Whooping, Mississippi sandhill, Cuban sandhill, Japanese, Siberian, hooded, white-naped, and black-necked cranes.)

Klataske, Ronald. 1972. Wings across the Platte. *National Wildlife* 10(5): 44–47.

Krapu, Gary L. 1987. Sandhill recovery. *Birder's World* 1(1):4–8. (Population trends in sandhill cranes.)

Krapu, Gary L. 1987. Use of staging areas by sandhill cranes in the midcontinent region of North America. In Archibald and Pasquier, 1987, pp. 451–462.

Kuyt, Ernie. 1987. Whooping crane migration studies, 1981–82. In Archibald and Pasquier, 1987, pp. 371–379.

Kuyt, Ernie, and P. Oossen. 1987. Survival, sex ratio, and age at first breeding of whooping cranes in Wood Buffalo National Park, Canada. In Lewis, 1987, pp. 230–244.

Lewis, James C. 1974. Ecology of sandhill cranes in the southeastern central flyway. Ph.D. dissertation, Oklahoma State University, Stillwater.

Lewis, James C. (ed.). 1976. Proceedings of the 1973 International Crane Workshop, Baraboo, Wisconsin. Stillwater: Oklahoma State University Publishing & Printing Dept. (Out of print.)

Lewis, James C. 1977. Sandhill crane. In G. C. Sanderson (ed.), Management of Migratory Shore and Upland Game Birds in North America, pp. 5–44. Washington, D.C.: International Association of Fish and Wildlife Agencies. Reprinted by University of Nebraska Press, Lincoln.

Lewis, James C. (ed.). 1979. Proceedings of the 1978 Crane Workshop, Rockport, Texas. Fort Collins: Colorado State University Printing Service. (Available from the National Audubon Society, 4150 Darley St., Suite 5, Boulder, Colo. 80303, $6.00.)

Lewis, James C. (ed.). 1982. Proceedings of the 1981 Crane Workshop. Tavernier, Florida: National Audubon Society. (Available from the National Audubon Society, 115 Indian Mound Trail, Tavernier, Fla. 33070, $25.00.)

Lewis, James C. (ed.). 1987. Proceedings of the 1985 Crane Workshop. Grand Island, Nebraska: Platte River Whooping Crane Habitat Maintenance Trust and U.S. Fish and Wildlife Service. (Out of print.)

Lewis, James C., and Hiroyuki Masatomi (eds.). 1981. Crane Research Around the World. Baraboo, Wisconsin: International Crane Foundation. (Proceedings of International Crane Symposium, Sapporo, Japan, 1980; available from the ICF, Baraboo, Wisconsin 53913, $17.00.)

McCoy, J. J. 1966. The Hunt for the Whooping Cranes: A Natural History Detective Story. New York: Lothrop, Lee & Shepard Co.

Mackenzie, John P. 1977. Birds in Peril. Boston: Houghton Mifflin. (Whooping crane.)

McMillen, J. L. 1988. Conservation of North American cranes. *American Birds* 42:1212–1221.

McNulty, Faith. 1966. The Whooping Crane: The Bird That Defies Extinction. New York: E. P. Dutton.

Madson, John. 1974. Day of the crane. *Audubon* 74(2):46–63. (Sandhill cranes and the Platte River.)

Masatomi, Hiroyuki. 1989. International censuses on wintering cranes in East Asia, 1987–88. International Crane Research Unit in Eastern Asia, Bibai, Japan.

Nesbitt, S. A., and A. S. Wenner. 1987. Pair formation and mate fidelity in sandhill cranes. In Lewis, 1987, pp. 117–122.

Nesbitt, S. A., A. S. Wenner, and J. H. Hintermister V. 1987. Progress of sandhill crane studies in Florida. In Archibald and Pasquier, 1987, pp. 411–414.

Olsen, D. L., D. R. Blankinship, R. C. Erickson, R. Drewien, H. D. Irby, R. Lock, and L. S. Smith. 1980. Whooping crane recovery plan. Washington, D.C.: U.S. Fish and Wildlife Service.

Reed, Jonathan R. 1988. Arctic adaptations in the breeding biology of sandhill cranes, *Grus canadensis*, on Banks Island, Northwest Territories. *Canadian Field-Naturalist* 102:643–648.

Reeves, Henry R. 1975. A contribution to an annotated bibliography of North American cranes, rails, woodcock, snipe, doves and pigeons. Washington, D.C.: U.S. Fish and Wildlife Service. (Distributed by National Technical Information Service, U.S. Dept. of Commerce; ref. code PB-240 999; survey of literature through 1971, with approximately 600 crane references.)

Safina, C., L. Rosenbluth, C. Pustmueller, K. Strom, R. Klataske, M. Lee, and J. Beyea. 1989. Threats to wildlife and the Platte River. Environmental Policy Analysis Department Report no. 33. New York: National Audubon Society.

Sherwood, Glen. 1971. If it's big and flies—shoot it. *Audubon Magazine* 73 (Nov.):72–99.

Shoemaker, T. G. 1988. Wildlife and water projects on the Platte River. In W. J. Chandler (ed.), Audubon Wildlife Report 1988/1989, pp. 285–334. San Diego: Harcourt, Brace, Jovanovich.

Sidle, John G. 1989. A prairie river roost. *Living Bird Quarterly* 8(2):8–13. (Sandhill cranes and the Platte River.)

Temple, Stanley A. 1978. Endangered Birds: Management Techniques for Preserving Threatened Species. Madison: University of Wisconsin Press. (Cross-fostering of whooping cranes.)

Turner, Tom. 1989. Trouble at Aransas. *Defenders* 64(3):30–34. (Threats to Aransas National Wildlife Refuge from the Intracoastal Waterway.)

# REFERENCES

U.S. Fish and Wildlife Service. 1981. The Platte River Ecology Study: Special Research Report. Washington, D.C.: U.S. Dept. of Interior.

U.S. Fish and Wildlife Service. 1986. Endangered and Threatened Wildlife and Plants. January 1. 1986. Washington, D.C.: Government Printing Office.

Vogt, William. 1978. Now, the river is dying. *National Wildlife* 16(4):4–11. (Platte River.)

Walkinshaw, Lawrence H. 1949. The Sandhill Cranes. Bloomfield Hills: Cranbrook Institute of Science Bulletin 29:1–202.

Walkinshaw, Lawrence H. 1981. Cranes of the World. New York: Winchester Press.

Walkinshaw, Lawrence H. 1981. Cranes of the world: a partial bibliography. In Lewis and Masatomi, 1981, pp. 24–45. (Over 900 supplementary references to the more than 1,500 citations provided in Walkinshaw's 1981 book.)

Walkinshaw, L. H. 1986. The Sandhill Crane and I. Ann Arbor: University Microfilms, no. LD01109. (History of the sandhill crane in Michigan.)

Walkinshaw, L. H. 1987. Nesting of the Florida and Cuban Sandhill Cranes. Ann Arbor: Univ. Microfilms, no. LD01165. 378 pp.

Zimmerman, Dale R. 1975. To Save a Bird in Peril. New York: Coward, McCann & Geoghegan.

Zimmerman, Dale R. 1978. A technique called cross-fostering may help save the whooping crane. *Smithsonian* 9(6):52–63.

Zimmerman, Dale R. 1981. A fragile victory for beauty on an old Asian battleground. *Smithsonian* 12(7):57–65. (Crane conservation efforts in Korea.)

Additionally, the nonprofit International Crane Foundation, of Baraboo, Wisconsin, publishes a popularly written and beautifully illustrated quarterly newsletter for its members, titled *The Bugle*. The ICF also has helped sponsor and publish the proceedings of several workshops on crane research around the world (see, for example, Archibald and Pasquier, 1987), has played a major role in developing avicultural techniques for breeding rare cranes in captivity, and has promoted the conservation of all species of cranes worldwide (see Harrison, 1978). Information on membership in the ICF and its activities can be obtained from the International Crane Foundation, E-11376 Shady Lane Road, Baraboo, WI 53913. Likewise, the North American Crane Working Group publishes information for its members on news and research related to the North American cranes. Its mailing address is 2550 N. Diers Ave., Suite H, Grand Island, NE 68803.

# INDEX